D0077801

STUDENT WORKBOOK

LINDA DAWSON
University of Washington

THE TRIOLA STATISTICS SERIES

Mario F. Triola
Dutchess Community College

PEARSON

Boston   Columbus   Indianapolis   New York   San Francisco   Upper Saddle River
Amsterdam   Cape Town   Dubai   London   Madrid   Milan   Munich   Paris   Montreal   Toronto
Delhi   Mexico City   São Paulo   Sydney   Hong Kong   Seoul   Singapore   Taipei   Tokyo

The author and publisher of this book have used their best efforts in preparing this book. These efforts include the development, research, and testing of the theories and programs to determine their effectiveness. The author and publisher make no warranty of any kind, expressed or implied, with regard to these programs or the documentation contained in this book. The author and publisher shall not be liable in any event for incidental or consequential damages in connection with, or arising out of, the furnishing, performance, or use of these programs.

Reproduced by Pearson from electronic files supplied by the author.

Copyright © 2014, 2012, 2008 Pearson Education, Inc.
Publishing as Pearson, 75 Arlington Street, Boston, MA 02116.

All rights reserved. No part of this publication may be reproduced, stored in a retrieval system, or transmitted, in any form or by any means, electronic, mechanical, photocopying, recording, or otherwise, without the prior written permission of the publisher. Printed in the United States of America.

ISBN-13: 978-0-321-89196-9
ISBN-10: 0-321-89196-1

6

www.pearsonhighered.com

PEARSON

Table of Contents

CHAPTER 1: INTRODUCTION TO STATISTICS

EXAMPLE 1: Populations and Samples

Name a population of which you are a part. From that population, describe a sample of which you are a part. Then, describe a sample of which you are not a part.

Possible Response #1

The collection of all members of the Society of Women Engineers (SWE) is a population. The author is a member of that population.

- A sample from that population that includes the author is the collection of all SWE members who are aerospace engineers.

- A sample from that population that does not include the author is the collection of all SWE members who are aerospace engineers who live in California.

Possible Response #2

The collection of all basketball players in the NBA is a population. Suppose you are Kobe Bryant answering the question.

- A sample from that population that contains Kobe Bryant is the collection of all NBA players on the LA Lakers team.

- A sample from that population that does not contain Kobe Bryant is the collection of all NBA players on the Boston Celtics.

EXAMPLE 2: Populations and Samples

During the 2009-2010 school year, 2,800 kids and teens at 20 randomly selected middle and high schools in the Minneapolis/St. Paul area completed a survey on food and weight-related behaviors, including activities tied to muscle gain. In addition to steroid use, more than one-third of boys and one-fifth of girls in the study said they had used protein powder or shakes to gain muscle mass, and between five and 10 percent used non-steroid muscle-enhancing substances.

What is the population represented in this survey? What is the sample?

- The population is the larger group being studied. In this example the large group is middle and high school students in the Minneapolis/St. Paul area.

- The sample is the collection from the population who provide the actual data. In this example, the sample is the 2800 students from randomly selected schools who responded to the survey.

EXAMPLE 3: Levels of Measurement

Consider the population of all currently licensed Chevrolet vehicles on American highways.

1. Give an example of data from this population at the nominal level of measurement.

- The nominal level of measurements has categories with no ordering scheme.

- Organize the Chevrolet vehicles into categories such as sedans, minivans, trucks, SUVs, 4-wheel drives, and so on.

2. Give an example of data from this population at the ratio level of measurement.

- The ratio level of measurement is numerical with a meaningful zero value and meaningful ratio values.

- Current blue book resale value of each Chevrolet vehicle is an example of a ratio level of measurement.

- The odometer reading on each vehicle is another example of a ratio level of measurement.

VOCABULARY CHECK

1. Give an example of a voluntary response study. (Try to find one that is not described in your textbook!)

2. Give an example of an observational study. (Try to find one that is not described in your textbook!)

3. Give an example of an experimental study. (Try to find one that is not described in your textbook!)

4. Give an example of a stratified sample taken from the population of American college students.

5. Give an example of a convenience sample taken from the population of teenage drivers.

6. Give an example of a systematic sample taken from the population of shoppers at a major department store.

7. A numerical measure describing some feature of a population is called a _____; a numerical measure describing some feature of a sample is called a _____.

8. Give an example of a type of data at the nominal level of measurement.

9. Give an example of statistical significance versus practical significance. (Try to find one that is not described in your textbook!)

Name: Date:
Instructor: Section:

SHORT ANSWER

1. Name three reasons for gathering data from a sample instead of from the entire population.

2. In 2012, third baseman Miguel Cabrera of the Detroit Tigers baseball team had 205 hits in 622 times at bat. What is Miguel's batting average?

3. In a recent survey of 170 sociology students at the University of North Florida, 52.4% of the males strongly agreed that a married woman should take her husband's last name. How many male sociology students does this 52.4% represent?

4. TRUE or FALSE: Correlation implies causality. Defend your answer with a brief explanation.

5. TRUE or FALSE: Very large samples guarantee sound statistical results. Defend your answer with a brief explanation.

6. TRUE or FALSE: A sample of 1250 responders to an online survey is an example of a self-selected sample. Defend your answer with a brief explanation.

Copyright ©2014 Pearson Education, Inc.

CHAPTER 2: SUMMARIZING AND GRAPHING DATA

EXAMPLE 1: Constructing Various Displays of Data

The number of college credits completed by a sample of 53 Elementary Statistics students is shown below. Use this information to respond to the questions that follow.

9 9 9 12 12 12 12 18 18 18 19 20 27 30 30 33 35 35 37 39 39 42 43 43 45 47 50 50 52 53 56 57 57 57 60 64 65 66 70 72 73 76 76 80 84 90 92 103 106 109 109 120 120

1. We want to create a frequency distribution with five classes. What class width should we use?

$$\text{Class width} = \frac{\text{high data value - low data value}}{\text{total number of classes}} = \frac{120-9}{5} = \frac{111}{5} = 22.5 \rightarrow 23$$

Since our data values are whole numbers, our class width should also be a whole number. We round up to the next highest whole number.   Class width = 23.

2. What are the class limits of the five classes in our frequency distribution?

The lower class limit of the first class is the lowest data value. Use the class width to compute the class limits of each new class.

classes	frequency
9	
9+23=32	
32+23=55	
55+23=78	
78+23=101	

We fill in the upper class limits by making sure that the classes do not overlap and that the classes have no gaps between them.

classes	Frequency
9-31	
32-54	
55-77	
78-100	
101-123	

Notice that the upper class limits are separated by the class width value of 23, just as the lower class limits were. Also, notice that our highest data value belongs to the last class while our lowest data value belongs to the first class.

Copyright ©2014 Pearson Education, Inc.

3. Find the frequency corresponding to each class.

Using the same kind of tally marks you used in grade school, slot each data value into the appropriate class and then find the total for each class.

Hint: you may find this easier to do by first putting your data in order (the calculator does this very efficiently). You can then simply count the number of values in each class.

class	frequency
9-31	15
32-54	15
55-77	13
78-100	4
101-123	6

4. Find the midpoint for each class. Midpoint = $\dfrac{\text{upper class limit} + \text{lower class limit}}{2}$

Class	frequency	midpoint
9-31	15	(31+9)/2=20
32-54	15	(54+32)/2=43
55-77	13	(77+55)/2=66
78-100	4	(100+78)/2=89
101-123	6	(123+101)/2=112

Again we notice that the midpoints are separated by a distance equal to the class width.

5. Find the class boundaries for each class.

The upper class boundary of the first class and the lower class boundary of the second class are the same. They correspond to the value that is "in between" the upper class limit of one class and the lower class limit of the next class.

Lower class boundary	class	Upper class boundary
	9-31	(31+32)/2=31.5
(31+32)/2=31.5	32-54	(54+55)/2=54.5
(54+55)/2=54.5	55-77	(77+78)/2=77.5
(77+78)/2=77.5	78-100	(100+101)/2=100.5
(100+101)/2=100.5	101-123	

Notice once again, that the class boundaries are separated by an amount equal to the class width. We can use this observation to complete the first lower class boundary and last upper class boundary.

Lower class boundary	class	Upper class boundary
31.5-23=8.5	9-31	31.5
31.5	32-54	54.5
54.5	55-77	77.5
77.5	78-100	100.5
100.5	101-123	100.5+23=123.5

6. Construct a relative frequency distribution for this data.

In each class, the relative frequency corresponds to the percent of the total data that falls in that class. This percent is expressed in decimal form rounded to three places.

$$\text{Relative frequency} = \frac{\text{class frequency}}{\text{total number of data values}}$$

In this data set we have 53 data values so in each calculation our denominator is 53.

class	frequency	Relative frequency
9-31	15	15/53= 0.283 = 28.3%
32-54	15	15/53= 0.283 = 28.3%
55-77	13	13/53= 0.245 = 24.5%
78-100	4	4/53= 0.075 = 7.5%
101-123	6	6/53= 0.113 = 11.3%

Notice that the sum of the fractional parts is exactly 1 while the sum of the percentages is 99.9%. The missing 0.1% is due to our round-off error.

Copyright ©2014 Pearson Education, Inc.

7. Construct a cumulative frequency distribution for this data.

In this table, we have only upper class boundaries. Our frequencies "accumulate" as we move to each new class. In the last class, the cumulative frequency must equal our sample size.

class	Cumulative frequency
Less than 31.5	15 (frequency of first class)
Less than 54.5	15+15 = 30 (sum of frequencies of first two classes)
Less than 77.5	15+15+13 = 43 (sum of frequencies of first three classes)
Less than 100.5	15+15+13+4 = 47 (sum of frequencies of first four classes)
Less than 123.5	15+15+13+4+6= 53 (sum of frequencies of all five classes)

8. Construct a frequency polygon for this data.

In a frequency polygon we "connect the dots."
  • For each class, in the distribution we create an ordered pair that looks like: (midpoint of the class, frequency for the class).

  • Along the horizontal axis of your graph mark off units corresponding to class midpoints.

  • Along the vertical axis, mark off units corresponding to class frequencies.

  • Use a scale that shows enough detail to get a feeling for the differences between the classes.

  • Do not use a scale that has a large amount of empty space above the graph or to the left and right along the horizontal axis.

  • Extend both ends of the graph down to the x-axis (this creates the closed "polygon"). To find the appropriate x-value, add the class width to the final class midpoint and then subtract the class width from the initial class midpoint (20 – 23 = -3 and 112 + 23 = 135).

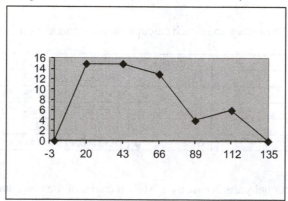

9. Construct a histogram for this data.

The histogram looks like a bar graph with some very specific requirements.
- Along the horizontal axis of your graph, mark off units corresponding to class boundaries.
- Along the vertical axis of your graph mark off units corresponding to frequencies.
- Create vertical bars for each class: the bar should cover the interval from its lower class boundary to its upper class boundary and should be as tall as its frequency.
- The bars in your histogram will not overlap but they will touch each other.

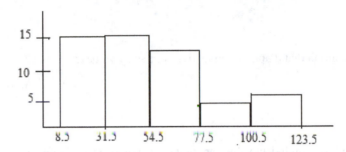

10. Create an ogive for this data.

Once again we connect the dots.
- For each class we create an ordered pair in the form
  (upper class boundary, cumulative frequency for the class).
- Label the horizontal axis with the class boundaries.
- Label the vertical axis with the cumulative frequencies.
- Notice that this graph will always increase; a segment may be horizontal (corresponding to some class with a frequency of 0) but it can never decrease because we never lose any of the accumulated frequencies.

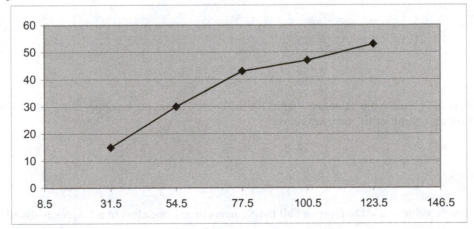

Copyright ©2014 Pearson Education, Inc.

VOCABULARY AND SYMBOL CHECK

1. What do we call the highest value that is possible to be put in any class of a frequency distribution?

2. What do we call a bar graph for qualitative data?

3. What do we call a line graph that displays cumulative frequency values?

4. What do we call a graph in which data values are plotted as points or dots along a scale of values?

5. What do we call a graph of pairs of data values using a horizontal and a vertical axis?

6. The mnemonic CVDOT is used to help remember the important characteristics of data. What does CVDOT stand for?

7. What is the name given to the value that "splits the difference" between the upper class limit of one class and the lower class limit of the very next class?

8. What one single value is used to represent all the elements in any one class of a frequency distribution?

Copyright ©2014 Pearson Education, Inc.

SHORT ANSWER

1. What is the formula for finding the class midpoint?

2. Identify two important characteristics of a normal or bell-shaped distribution.

3. Identify two details that might make a graph misleading or inaccurate.

4. In a time-series graph, the horizontal axis is marked off in units of _____?

5. Name one guaranteed characteristic of an ogive.

6. What can be gained by presenting data in a frequency distribution? What is lost by this presentation?

Copyright ©2014 Pearson Education, Inc.

7. What can be gained by presenting data in a relative frequency distribution? What is lost by this presentation?

8. TRUE or FALSE: We may adjust the class limits of a frequency distribution to convenient or attractive values even if this causes the first or last class to be empty. Defend your answer.

9. TRUE or FALSE: If some class other than the first or last class of a frequency distribution is empty, we must adjust the class limits so that there is at least one data value in the class. Defend your answer.

10. TRUE or FALSE: The classes in a frequency distribution may overlap in order to make all the data fit. Defend your answer.

11. TRUE or FALSE: To find class width, subtract the lower class limit from the upper class limit. Defend your answer.

12. Identify the differences in the following bar charts comparing bubble sizes created with 3 types of soap. Identify any misconceptions that can be created by presenting the data in the form on the right.

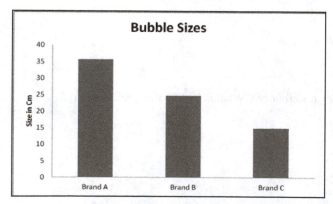
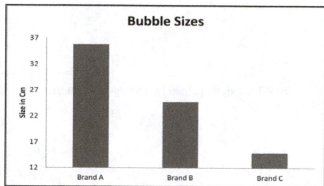

Copyright ©2014 Pearson Education, Inc.

Name: Date:
Instructor: Section:

PRACTICE PROBLEMS

1. What is the class width for the frequency distribution of a data set with the high data value of 625, a low data value of 52, and 10 classes?

2. What is the midpoint of a class with lower class limit 12 and upper class limit 66?

3. Find the relative frequency for a class with a frequency of 124 in a frequency distribution with $n = 500$.

4. Construct the stem-and-leaf plot for these times in a 5-K foot race. (Times are in minutes.)
   12, 23, 10, 24, 17, 32, 34, 41, 19, 25, 17, 24, 26, 31, 20, 21, 32, 19, 26, 29, 30, 40, 33, 28, 21

Copyright ©2014 Pearson Education, Inc.

Name: Date:
Instructor: Section:

5. Create the cumulative frequency distribution for this table

Miles in daily commute from home to work	Frequency
1-10	6
11-20	12
21-30	15
31-40	8
41-50	3
51-60	2

Classes	Cumulative Frequencies

Copyright ©2014 Pearson Education, Inc.

6. The data below is the number of states visited by students in a community college Chemistry class. Construct a frequency distribution having 5 classes for this data.

3 7 15 2 10 5 5 30 6 10 30 20 15 21 36 36 2 10 9 31 8 34 7 4 20 5 20 4 38 13

Classes	Frequency

Copyright ©2014 Pearson Education, Inc.

CHAPTER 3: STATISTICS FOR DESCRIBING, EXPLORING, AND COMPARING DATA

EXAMPLE 1: Finding Measures of Center from a Data List

The top 20 NFL all time game scores are:
72, 70, 65, 65, 64, 63, 63, 62, 62, 62, 62, 62, 62, 62, 61, 61, 61, 61, 60, 60. (Source: NFL.com, stats)

1. What is the mean number of all time game scores?

    Use the formula $\bar{x} = \dfrac{\sum x}{n}$ to find the sum of the data values and divide by that total number of values

    $$\frac{72+70+65+65+64+63+63+62+62+62+62+62+62+62+61+61+61+61+60+60}{20} = 63.00$$

    Round this number to one decimal place more than is present in the original set of values so $\bar{x} = 63.0$

2. What is the median number of completed passes?

    To find the median, arrange the data values in order from smallest to largest and find the value that is in the middle position of this list.

    60 60 61 61 61 61 62 62 62 62 62 62 62 63 63 64 65 65 70 72

    There are 20 data values so the middle position will be between the tenth and eleventh positions. The median is the arithmetic mean of 62 and 62: median $= \dfrac{62+62}{2} = 62.0$

3. What is the mode of this data set?

    The mode is the data value that is repeated most often. The value of 62 is repeated seven times and is repeated the most so is the mode of this data set.

4. What is the midrange of this data set?

    The midrange is the arithmetic mean of the high and low values in the set so we find

    Midrange $= \dfrac{60+72}{2} = 66.0$

Copyright ©2014 Pearson Education, Inc.

EXAMPLE 2: Finding Measures of Center from a Frequency Distribution.

Students in a physical education class were asked how many hours per week they spent working out in a gym. The table below summarizes the responses.

Hours spent in the gym	Frequency
0-4	10
5-9	12
10-14	7
15-19	4
20-24	2

1.   What is the mean number of hours spent in the gym?

Since we do not know the individual data values, we must substitute the class midpoint in place of each value in that class. Recall that the class midpoint is the mean value of each class.

Class midpoints: $\dfrac{0+4}{2} = 2$, $\dfrac{5+9}{2} = 7$, $\dfrac{10+14}{2} = 12$, $\dfrac{15+19}{2} = 17$, $\dfrac{20+24}{2} = 22$

Next we use the formula $\bar{x} = \dfrac{\sum(x \cdot f)}{\sum f}$ where x is the class midpoint and f is the corresponding class frequency.

$\bar{x} = \dfrac{(2)(10) + (7)(12) + (12)(7) + (17)(4) + (22)(2)}{35} = 8.57... \rightarrow 8.6$ (round to one place more than the raw data)

2.   What is the median number of hours spent in the gym?

The median is the data value in the middle position.

- Once again, we do not have raw data values so we use class midpoints.

- There are 35 data values (sum of the frequencies) so the middle position is the 18th position..

- The value in the 18th position falls in the second class (the first 10 values are in the first class, the next 12 values are in the second class) so the median is the midpoint of the second class.

- Median = 7.

3.   What is the mode of this data?
- The mode is the class midpoint of the class with the highest frequency.
- The second class has the highest frequency so the mode is the midpoint of the second class.
- Mode = 7

4.   What is the midrange of this data?
- The midrange is the arithmetic mean of the midpoints of the first and last classes.

- Midrange = $\dfrac{2+22}{2} = 12$

Copyright ©2014 Pearson Education, Inc.

EXAMPLE 3: Finding Measures of Variation.

Find the range, variance and standard deviation for the NFL highest scores data in Example 1 above.
72, 70, 65, 65, 64, 63, 63, 62, 62, 62, 62, 62, 62, 62, 61, 61, 61, 61, 60, 60

To find the range:

- Subtract the lowest data value from the highest.

- range $= 72 - 60 = 12.0$ (result shown with one more decimal place than is present in the original data.)

To find the variance we must apply the formula $s^2 = \dfrac{n\sum x^2 - (\sum x)^2}{n(n-1)}$ where $n = 20$ and x represents the individual data values.

- $\sum (x^2) = 72^2 + 70^2 + 65^2 + 65^2 + 64^2 + 63^2 + 63^2 + 62^2 + 7(62^2) + 4(61^2) + 60^2 + 60^2 = 79{,}560$

- $(\sum x)^2 = (\,72 + 70 + 65 + 65 + 64 + 63 + 63 + 62 + 62 + 62 + 62 + 62 + 62 + 61 + 61 + 61 + 61 + 60 + 60\,)^2$

  $= 1260^2 = 1{,}587{,}600$

- $n\sum(x^2) = 20(79{,}560) = 1{,}591{,}200$

- $n\sum(x^2) - (\sum x)^2 = 1{,}591{,}200 - 1{,}587{,}600 = 3600$

- $n(n-1) = (20)(19) = 380$

- $s^2 = \dfrac{3600}{(380)} = 9.474 \rightarrow 9.5$ (following our established round off rules)

To find standard deviation, take the square root of the variance: $s = \sqrt{s^2} = \sqrt{9.5} = 3.1$

EXAMPLE 4: Using the TI Calculator to Find Measures of Center and Measures of Variation.

Use your calculator to find the mean, median, mode, midrange, range, and standard deviation of the NFL highest score data from Example 1.

- Enter the data into a list.

    o The keystroke are   STAT   1:EDIT followed by ENTER

    o The list columns are now on the homescreen

    o Choose a convenient list and type in each data value

- Sort the data in your list

    o The keystrokes are: STAT   2:SortA(ENTER

    o You will now see   SortA(followed by a blinking cursor on your homescreen.

    o Type in the name of the list that has the data followed by ENTER

    o The data in your list is now in order from smallest to largest.

    o If you want the list ordered form largest to smallest, choose SortD(

- Find the statistic values

    o The keystrokes are: STAT, use right arrow to choose CALC 1:1-Var Stats ENTER

    o You will now see 1-Var Stats followed by a blinking cursor on your homescreen

    o Type in the name of the list that has the data followed by ENTER

- Read the display of statistics

    o $\bar{x} = 63$

    o $\sum x = 1260$ which is one of the values used in the computation of variance

Copyright ©2014 Pearson Education, Inc.

- $\circ$ $\sum x^2 = 79{,}560$ which is one of the values used in the computation of variance.

- $\circ$ $s_x = 3.077935$ which we round to 3.1. This is sample standard deviation.

- $\circ$ $\sigma_x = 3$. This is the population standard deviation.

Remember, the calculator does not know if the data in the list represents the entire population or is just a sample from some larger population, so it will compute both values for standard deviation, σ_x and s_x. It is up to you as chief-statistician-in-charge to select the correct value.

- Now use the down arrow to scroll down for the rest of the information.

  - $\circ$ $n = 20$ is the sample size

  - $\circ$ minX = 60 identifies the smallest data value

  - $\circ$ $Q_1 = 61$ identifies the value for the first quartile

  - $\circ$ Med = 62 identifies the median value (also called Q_2 for second quartile)

  - $\circ$ $Q_3 = 63.5$ identifies the value for the third quartile

  - $\circ$ maxX = 72 identifies the largest data value

  This is the 5-number summary for the NFL highest score data.

- To find the mode, you must scroll through the data in the list to identify any repeated values. As before, we observe no repeated values so this data set has no mode.

- To find the range and midrange, use the same formulas as in Example 1. The calculator has located minX and maxX for you.

Copyright ©2014 Pearson Education, Inc.

Name: _____   Date: _____

Instructor: _____   Section: _____

VOCABULARY AND SYMBOL CHECK

Fill in the blank with the most appropriate word or simple phrase.

1. The symbol for population mean is _____.

2. The symbol for sample mean is _____.

3. The symbol for population standard deviation is _____.

4. The symbol for sample standard deviation is _____.

5. The values included in the 5-number summary of a data set are _____.

6. The measure that describes how many standard deviations away from the mean a data value lies is called

    _____.

7. A percentile measures location by _____.

8. The formula for a z- score is _____.

9. What is the "range rule of thumb?"

10. What is the interquartile range?

Copyright ©2014 Pearson Education, Inc.

SHORT ANSWER

1. According to the empirical rule for symmetric distributions, how much of the data lies within two standard deviations of the mean?

2. A measure of center that is not affected by extreme values is _____.

3. A measure of center that may not be unique is _____.

4. To construct a boxplot for a data set you must first find _____.

5. How do we identify "unusual values" in a data set using the range rule of thumb?

6. The selling prices of a sample of homes in a particular neighborhood are gathered. There is one extremely high value in this set (a high outlier).

    a. Which measures of center would be the most affected by this extreme value?

    b. Which measures of center would be least affected by this extreme value?

Name: Date:
Instructor: Section:

7. TRUE or FALSE: The standard deviation of a set of data values is never a negative number. Defend your answer.

8. TRUE or FALSE: In a data set with one really extreme value, the mean would be a good choice for a reliable measure of central tendency. Defend your answer.

9. TRUE or FALSE: The variance is the square root of the standard deviation. Defend your answer.

10. TRUE or FALSE: The modal class of a grouped frequency distribution is the class that has the highest frequency. Defend your answer.

11. Round off rules:

   a. Round population parameter values to _____.

   b. Round sample statistic values to _____.

   c. Round z- scores to _____.

12 Explain why there is no 100th-percentile.

13. Under what circumstances would we use the coefficient of variation to compare variation in different data sets?

14. Choose all that apply: A data set may have outliers because:

   a. data entry error when data was recorded (typographical error).

   b. error in gathering data (incorrect measurement, incorrect instrument reading, faulty instrument, etc.).

   c. person responding to a survey question misunderstood the question.

   d. person responding to a survey question intentionally gave a bad answer.

   e. a data set has a legitimate unusual value.

Copyright ©2014 Pearson Education, Inc.

PRACTICE PROBLEMS

1. The number of hours of television watched per day by a sample of 28 people is given below.

    4 1 5 5 2 5 4 4 2 3 6 8 3 5 2 0 3 5 9 4 5 2 1 3 4 7 2 9

    a. Find the 5-number summary for this data set.

    b. Draw a reasonably accurate sketch of the box and whisker plot for this data. Remember to begin with a number line marked with units that correspond to your data values.

    c. What percent of the people watched more than four hours of television per day?

    d. About 75% of the people watch no more than how many hours of television per day?

Copyright ©2014 Pearson Education, Inc.

2.   The z- scores for eight students' statistics quiz grades are listed here:

   Adam: 1.22   Bob: 0.13   Carlos: -2.31   Demetrio: -0.45
   Ellen: 0.00   Fran: -1.65   Gert: 2.14   Harish: 0.67

   a.   Which, if any, of these students had quiz grades above the mean? Defend your answer.

   b.   Which, if any, of these students have quiz grades below the mean? Defend your answer.

   c.   Which, if any, of these students have quiz grades equal to the mean? Defend your answer.

   d.   Which, if any, of these students have unusually high quiz grades? Defend your answer.

   e.   Which, if any, of these students have unusually low quiz grades? Defend your answer.

Copyright ©2014 Pearson Education, Inc.

3. Chris scores 80 on a biology quiz that has a mean of 75 and a standard deviation of 3 and then scores 12 on a Spanish quiz that has a mean of 10 and a standard deviation of 1.1. Which quiz result is relatively better? Why?

4. Compare the variation in the age of adult women with variation in the weight of the women. A random sample of 15 women was used to gather the data.

Age: 19 22 21 21 20 19 19 21 46 19 24 34 21 38 20
Weight: 140 132 195 150 127 120 140 140 100 150 140 200 110 190 145

5. Listed below are the total credit hours completed by a sample of the author's students in a recent summer school Calculus I class. Find the mean, median, mode, midrange, range, and standard deviation of this data set. Which, if any of these results would be considered unusually high or unusually low? Defend your answer.   83 30 150 29 51 18 36 25 29 53 26 90 53 80 80

CHAPTER 4: PROBABILITY

EXAMPLE 1: Finding a Sample Space and Using It to Compute Simple Probabilities.

Suppose we roll one fair 6-sided die with one hand and flip a fair 2-sided coin with the other hand.

An example of a simple event that might result from this procedure is getting a 6 on the die paired with heads on the coin.

Another example of a simple event is getting a 5 on the die paired with tails on the coin.

For the complete sample space we must pair every outcome from the roll of the die with every outcome from the coin flip. Let H represent heads and let T represent tails. The complete sample space is
{1-H, 1-T, 2-H, 2-T, 3-H, 3-T, 4-H, 4-T, 5-H, 5-T, 6-H, 6-T}

Use this sample space to compute the following:

1. P(getting 4 paired with heads) = $\dfrac{\text{Number of outcomes with 4-H}}{\text{size of sample space}} = \dfrac{1}{12}$

2. P(getting any even number paired with tails) = $\dfrac{\text{number of even numbers paired with tails}}{\text{size of sample space}} = \dfrac{3}{12} = \dfrac{1}{4}$

    Notice that the elements in the sample space that we want to count are 2-T, 4-T and 6-T. These are the three events that are in our numerator.

3. P(getting two heads) = 0
    This is an impossible event since we are not flipping two coins.

Copyright © 2014 Pearson Education, Inc.

Name: _____   Date: _____
Instructor: _____   Section: _____

EXAMPLE 2: Computing Probability From a Contingency Table.

In a survey of a recent Statistics class at a commuter college in Florida, students were asked if they owned their own home, rented an apartment or house, or lived with their parents. The table below shows the responses:

	Own Home	Rent	Live with Parents
Male	3	12	5
Female	2	7	9

1. How many students were surveyed in this class?

    *   Add up the entries in each cell: 3+12+5+2+7+9 = 38 total students.
    *   This is the sample size, so $n = 38$.

2. What is the probability of selecting one student from this class and getting a male who owns his own home?

    $$\frac{\text{number of male homeowners}}{\text{total number in sample}} = \frac{3}{38} = 0.0789 \text{ (Round decimals to three significant digits.)}$$

3. What is the probability of selecting one student from this class and getting a female?

    $$\frac{\text{total number of females}}{\text{total number in sample}} = \frac{18}{38} = 0.474 \text{ (Round decimals to three significant digits.)}$$

4. What is the probability of selecting one student from this class and getting a male **or** someone who is renting a house or apartment?

    $$\frac{\text{number of males + number of renters - number of male renters}}{\text{total number in sample}} = \frac{20 + 19 - 12}{38} = \frac{27}{38} = 0.711$$

    (Round decimals to three significant digits.)

    *   The word "**or**" in the question is our clue to use the addition rule.

    *   Notice that when we count the number of events that belong in the numerator we must be sure not to count those male renters twice – we can count them as males or we can count them as renters but we cannot include them in both counts – they are the same guys!

Copyright © 2014 Pearson Education, Inc.

5. What is the probability that one student is selected and that student is a renter given that the student is known to be a male?

$$\frac{\text{number of male renters}}{\text{total number of males}} = \frac{12}{20} = \frac{3}{5} = 0.6$$

- Notice that the sample space has changed because we know we have selected a male.

- The denominator is now the total number of males.

- The numerator counts only the males who are renters.

6. What is the probability of selecting one student from this class and getting a male or a female?

Since everybody in this sample is either male or female, we are describing a guaranteed or certain event so the probability is 1.

7. What is the probability of selecting two students from this class (without replacement) and getting two females?

- The "**and**" word and the act of selecting two students tells us to use the multiplication rule.

- P(selecting 2 females) = P(1st selection is female **and** 2nd selection is female)

= P(1st selection is female) $\times$ P(2nd selection is female) =

$$\frac{\text{number of females in first selection}}{\text{total number of students}} \times \frac{\text{number of females remaining in second selection}}{\text{total number of students remaining in second selection}} = \left(\frac{18}{38}\right)\left(\frac{17}{37}\right) = 0.218$$

Name: Date:

Instructor: Section:

VOCABULARY AND SYMBOL CHECK:

Fill in the blank with the most appropriate word or simple phrase.

1. The probability of a certain or guaranteed event is _____.

2. The probability of an impossible event is _____.

3. The symbol for sample size is _____.

4. The symbol for the complement of event A is _____.

5. The complement of "at least one" is _____.

6. Two events that cannot occur at the same time are called _____.

7. The complement of "at least 4 out of 10" is _____.

8. An arrangement of items in which order makes a difference is called a _____.

9. An arrangement of items in which order does not make a difference is called a _____.

10. When the occurrence of one event changes the probability of the occurrence of another event, then the

   events are called _____.

Copyright © 2014 Pearson Education, Inc.

SHORT ANSWER

1. $6! =$ _____.

2. $_{12}C_5 =$ _____.

3. $_{12}P_5 =$ _____.

4. If $P(A) = 0.435$, then $P(\overline{A}) =$ _____.

5. For each of the following, decide if the number is a legitimate probability value. In each case, defend your answer.

    a. 1/3

    b. 0.658

    c. - 0.459

    d. 1.54

    e. 1234/5678

6. TRUE or FALSE: The complement of "all boys" is "all girls." Defend your answer.

7. TRUE or FALSE: For all events A and B, $P(A \text{ or } B) = P(A) + P(B)$. Defend your answer.

8. TRUE or FALSE: The probability of an impossible event is 0. Defend your answer.

9. TRUE or FALSE: 1000 different three-digit security codes can be constructed if repetition of digits is allowed. Defend your answer.

10. TRUE or FALSE: 125 "words" can be constructed using only the vowels a, e, i, o, and u with no repetition of vowels permitted. Defend your answer

11. TRUE or FALSE: There are 720 different permutations of letters in the name TRIOLA. Defend your answer.

Copyright © 2014 Pearson Education, Inc.

PRACTICE PROBLEMS

1. How many different permutations are there of the letters in the name MARIO TRIOLA?

2. In a recent advertising campaign, Applebee's Restaurant offered diners a special 3-course menu for $15.99. The menu included choices of three appetizers (Mozzarella sticks, house salad, soup), four entrees (chicken, fish, beef, vegetarian pasta), and two desserts (ice cream, apple pie).

   a. How many different 3-course meals can be ordered from this menu? (HINT: Consider drawing a tree diagram to help visualize this sample space; or use the counting rule to find the total number of meals.)

   b. How many of these 3-course meals have the beef course as the entrée?

   c. How many of these 3-course meals do not have soup as the appetizer?

   d. What is the probability that a 3-course meal has the chicken entrée?

Copyright © 2014 Pearson Education, Inc.

e. What is the probability that a 3-course meal has either soup or salad?

f. What is the probability that a 3-course meal has either ice cream or apple pie for dessert?

g. What is the probability that a 3-course meal has shrimp cocktail as the appetizer?

3. A certain statistics class has 18 female students and 12 male students.

   a. What is the probability of selecting two students from this class without replacement and selecting 2 females?

   b. What is the probability of selecting two students from this class without replacement and selecting 2 males?

   c. What is the probability of selecting two students from this class without replacement and selecting one male and one female? (Be careful! Think of all the possibilities.)

Copyright © 2014 Pearson Education, Inc.

d. The professor in this class would like to organize a committee to help critique a new textbook.

    i. How many different 5-person committees can be formed?

    ii. How many different all male 5-person committees can be formed?

    iii. How many different all female 5-person committees can be formed?

    iv. Use your responses to parts (ii) and (iii) above to find how many 5-person committees with mixed gender representation can be formed. (That means not all male and not all female; think complements!)

CHAPTER 5: DISCRETE PROBABILITY DISTRIBUTIONS

EXAMPLE 1: Creating and Using a Discrete Probability Distribution.

Joe Gamer has invented a game played with an unusually shaped 5-sided die. Each face of the die has a numeral between 1 and 5 on it. He has rolled this die many times and recorded the results, shown in the table below.

Numeral on face	1	2	3	4	5
Frequency	25	62	12	56	45

1.   How many times did Joe roll the die?
     Add up the individual frequencies: 25+62+12+56+45 = 200

2.   Compute the relative frequency table for this data.

     Recall that relative frequency is found by computing the ratio $\dfrac{\text{class frequency}}{\text{total frequencies}}$ for each class.

Numeral on face	1	2	3	4	5
Relative Frequency	$\dfrac{25}{200} = .125$	$\dfrac{62}{200} = .310$	$\dfrac{12}{200} = .060$	$\dfrac{56}{200} = .280$	$\dfrac{45}{200} = .225$

3.   Now create the probability distribution: the value of the random variable x corresponds to the numeral on the face; the value of the corresponding probability is the relative frequency. The probability distribution is:

x	1	2	3	4	5
P(x)	.125	.310	.060	.280	.225

4.   Confirm that this is a legitimate probability distribution.
     a.   Are all possible values of the random variable listed?
          The variable x represents the numerals on the faces of the die. There are only 5 faces so $x = 1, 2, 3, 4, 5$ covers all possible values.
     b.   All the probability values are between 0 and 1.
     c.   The sum of the probability values is exactly 1.

Copyright ©2014 Pearson Education, Inc.

5. Find the mean and the standard deviation of this distribution.
   a. If performing this computation by hand, use the formulas in your text:

   - $\mu = \sum [(x)P(x)]$

   - Substitute to get $\mu = 1(.125) + 2(.310) + 3(.060) + 4(.280) + 5(.225) = 3.17$.

   - $\sigma = \sqrt{\sum [(x^2)P(x)] - \mu^2}$

   - Substitute to get
     $\sigma = \sqrt{[1^2(.125) + 2^2(.310) + 3^2(.060) + 4^2(.280) + 5^2(.225) - 3.17^2]} = 1.40$.

   b. If using your calculator to find these values:

   - Enter the x values into L1 (or some other convenient list).

   - Enter the $P(x)$ values into L2 (or some other convenient list).

   - Choose STAT   CALC   1:1-var Stats enter.

   - Type in the list with your x values, followed by the list with your $P(x)$ values (important to enter the name of the x list first, followed by a comma, followed by the name of the $P(x)$ list).

   - Enter

   Read $\bar{x} = 3.17$, $\sigma = 1.40$

6. Using the range rule of thumb, would any of the outcomes in this distribution be considered unusual?

   According to the range rule of thumb any outcome outside the range of $\mu \pm 2\sigma$ is unusual. For this distribution, $\mu + 2\sigma = 3.17 + 2(1.40) = 5.97$ and $\mu - 2\sigma = 0.37$. There are no x values greater than 5.97 and none less than 0.37 so none of the outcomes is considered unusual.

7. Use the probabilities to determine if any of these outcomes is unusual.

   An outcome is unusual if $P(x) \le 0.05$. None of the $P(x)$ values in the table is less than 0.05 so none of our outcomes is considered to be unusual.

Copyright ©2014 Pearson Education, Inc.

8. What is the probability that a player rolls this die three times and gets all 1's? (This is a very good result in Joe's new game.)

    - P (all 1's in three rolls) $= P$ (1st roll is a 1 **and** 2nd roll is a 1 **and** 3rd roll is a 1)

      The **and** connective tells us to use the multiplication rule.

    - $P(1) \times P(1) \times P(1) = (0.125)(0.125)(0.125) = 0.002$
      Since these rolls of the die are independent, this is a "with replacement" problem.

    - Since this probability is less than 0.05, we conclude that achieving this result is unusual.

EXAMPLE 2: Recognizing and Using a Binomial Distribution.

A survey of undergraduate calculus and physics students showed that 58% of them had donated blood at least once. What is the probability that in any undergraduate class of 30 calculus and physics students there are exactly 20 blood donors?

1. Confirm that this is a binomial distribution.

    - Fixed number of trials: there are 30 students in the class.

    - Two categories of outcome: success = blood donor; failure = not a blood donor.

    - Trial are independent: probability of one student being a donor does not affect probability of any other student being a blood donor.

    - Probability of success on any one trial stays the same from trial to trial: p(blood donor) = .58.

2. Identify the variables in this problem.

    - Number of trials: $n = 30$.

    - Probability of success on any one trial: $p = .58$.

    - Probability of failure on any one trial: $q = 1 - p = 1 - .58 = .42$.

    - Number of successes: x = 20.

3. Compute the probability of exactly 20 successes in 30 trials.

    a. Using the formula for binomial probability:

    $$P(x) = \frac{n!}{(n-x)!x!} \cdot p^x \cdot q^{n-x} \quad \text{using n = 30, x = 20, p = .58, q = .42 we get}$$

    $$P(x) = \frac{30!}{(30-20)!20!} \cdot \left(.58^{20}\right)\left(.42^{30-20}\right) = 0.095$$

    b. We cannot use Table A-1 to find this probability because $p = .58$ is not an option on the table.

    c. Using the calculator
        *   For the TI83 models: choose 2nd Vars 0:binompdf followed by ENTER
            For the TI84 models: choose 2nd Vars A:binompdf followed by ENTER

        *   on the homescreen you should see *binompdf(* followed by a blinking cursor

        *   type in 30, .58, 20 followed by ENTER (this corresponds to n, p, x)

        *   read 0.095

4. Find the mean and the standard deviation of this distribution.
    * $\mu = np$ so substitute to find $\mu = 30*.58 = 17.4$

    * $\sigma = \sqrt{npq}$ so substitute to get $\sigma = \sqrt{30*.58*.42} = 2.7$

5. What is the probability that at most 12 students in a class of 30 are blood donors?

    "At most 12" means "0 or 1 or 2 or … or 12" students are blood donors. Computing each of these probabilities without technology is extremely inefficient and even more tedious so we use the calculator.

    *   For the TI83 models: Choose 2nd Vars A:binomcdf followed by ENTER

        For the TI-84 models: Choose 2nd Vars B:binomcdf followed by ENTER

        The binomcdf option computes cumulative binomial probabilities from $x = 0$ up to and including the value you enter for x.

    *   On the homescreen you should see *binomdcf(* followed by a blinking cursor

    *   Type in 30, .58, 12 (this corresponds to getting up to and including 12 successes in 30 trials)

    *   Read 0.0358

Copyright ©2014 Pearson Education, Inc.

Name: Date:
Instructor: Section:

VOCABULARY AND SYMBOL CHECK

Fill in the blank with the most appropriate word or simple phrase.

1. In a binomial distribution, n represents _____

2. In a discrete probability distribution, x represents _____.

3. The formula for the mean of a binomial distribution is _____.

4. Suppose x represents the number of customers that a certain bank teller sees in an hour. Is x a continuous random variable or a discrete random variable? Why?

5. In a binomial distribution, p represents _____.

Copyright ©2014 Pearson Education, Inc.

SHORT ANSWER

1. In a binomial experiment with $p = .35$, P (exactly 5 successes in 7 trials) is _____.

2. In a binomial experiment, if $p = .46$, then q = _____.

3. In a binomial experiment, if $q = .248$, then p = _____.

4. In a Poisson distribution, if $n = 450$ and $p = .02$, then μ =_____.

5. What is the difference between "exactly 6 successes in 10 trials" and "at most 6 successes in 10 trials?"

6. TRUE or FALSE: In a probability distribution, the probability values may be any kind of number as long as the total of all the probabilities is exactly 1. Defend your answer.

7. TRUE or FALSE: In a binomial probability distribution, the probability of success will always be ½ since there are only two outcomes, success and failure. Defend your answer.

8. TRUE or FALSE: The mean of a probability distribution will always be one of the possible outcomes in the sample space for the experiment. Defend your answer.

9. TRUE or FALSE: An example of a binomial experiment is: spinning a roulette wheel 12 times and recording the number of times that the outcome is an odd number. Defend your answer.

10. TRUE or FALSE: An example of a binomial experiment is: spinning a roulette wheel 12 times and recording the number that comes up on each spin. Defend your answer.

Copyright ©2014 Pearson Education, Inc.

PRACTICE PROBLEMS

1. This table is a probability distribution in which x represents the number of students that a statistics tutor may see on any given day and $P(x)$ represents the probability that the tutor sees that number of students.

x	0	1	2	3	4	5
$P(x)$	0.11	0.18	0.36	0.16	0.14	0.05

    a. Confirm that this is a legitimate probability distribution by stating the conditions that must be satisfied and showing how they are satisfied.

    b. Find the mean and the standard deviation of this distribution.

    c. Based on this distribution, what is the probability that a tutor sees at least two students on a certain day?

Copyright ©2014 Pearson Education, Inc.

d.   Based on this distribution, what is the probability that a tutor sees either three or four students on a certain day?

e.   Based on this distribution, what is the probability that a tutor sees exactly 10 students on a certain day?

2.   A recent survey by the American Automobile Association revealed that 80% of teenage girls text while driving. You have been hired to do a safety presentation to a high school class of 100 teenage girls. You will ask how many of them text while driving.

a.   Explain why this procedure results in a binomial distribution.

b.   State the values of n, p and q for this distribution.

Copyright ©2014 Pearson Education, Inc.

c. What is the mean number of girls in this class who text while driving?

d. What is the standard deviation of this distribution?

e. What is the probability that exactly 75 of these girls text while driving?

f. What is the probability that at most 75 of these girls text while driving?

g. What is the probability that at least 75 of these girls text while driving?

Copyright ©2014 Pearson Education, Inc.

3. Use the Poisson distribution to find the probability of getting 2 successes in 500 trials when p=.015. Find the same probability using the binomial distribution. Compare your results and comment.

Copyright ©2014 Pearson Education, Inc.

CHAPTER 6: NORMAL PROBABLITY DISTRIBUTIONS

EXAMPLE 1: Finding Values in a Standard Normal Distribution.

1. Find $P(1 < z < 2)$

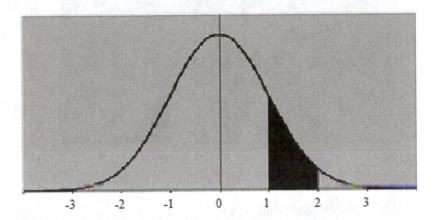

a.   Using technology

- Choose: 2nd VARS 2:normalcdf(followed by ENTER

- type: 1, 2) followed by ENTER

  This tells the calculator:

  o   the value of the left hand boundary of the shaded region

  o   the value of the right had boundary of the shaded region

  o   Since this is a standard normal distribution, there is no need to enter the values of m and s .

- Read: .1359

b.   Using Table A2:

We must find the area to the left of $z = 2$, then the area to the left of $z = 1$, then find the difference.

- Area to left of $z = 2$: down the z- column to 2.0, over to the first column headed by .00, read .9772.

- Area to left of $z = 1$: down the z- column to 1.0, over to the first column headed by .00, read .8413.

- Find the difference: $.9772 - .8413 = .1359$

Copyright ©2014 Pearson Education, Inc.

2. Find $P(1.34 < z)$

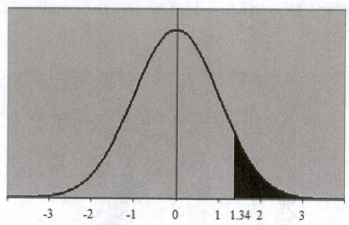

a.   Using technology

- Choose 2nd VARS   2:normalcdf(followed by ENTER

- Type: 1.34, 99999)

   Tell the calculator:

   o   the value of the left hand boundary of the shaded region.

   o   Since there is no right hand boundary (remember that the normal curve never actually reaches the horizontal axis), choose a very large value that is many standard deviations away from the mean. This guarantees that area value for our four decimal places will be accurate. (Experiment using a variety of "large" values for the right hand boundary and observe that all of them will produce the same area value.)

   o   Since this is a standard normal distribution there is no need to enter m and s .

- Read: .0901

b.   Using Table A2:

   We will find the area to the left of the shaded region, then subtract that value from 1. Recall that the total area under the normal curve is exactly 1.

- Go down the z- column to 1.3, then over to the first column headed by .04
   Notice that the z- value in question can be expressed as $1.3 + .04 = 1.34$

- Read .9099
   This is the total area to the left of $z = 1.3$.

- Subtract from 1 to get $1 - .9099 = 0.0901$
   This is the total area to the right of $z = 1.3$

Copyright ©2014 Pearson Education, Inc.

3. What z- value separates the bottom 20% of the area from the remaining area?
   Notice that the shaded area lies to the left of the mean so we expect a negative z- value for our answer.

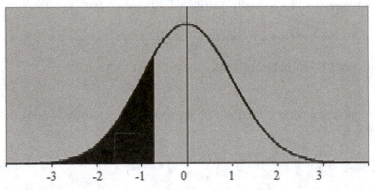

Shaded region corresponds to lowest 20% of the area.

a. Using technology:
   - Choose 2nd VARS 3:invNorm(followed by ENTER
     This is the function that "reverses" the normal distribution table in the calculator.

   - Type: .20) followed by ENTER
     Tell the calculator the total area to the left of the desired z-value.

   - Read: -.84 (rounded to two decimal places)
     The z- value is negative, as we expected.

b. Using Table A2
   - Since we have noted that our z- value will be negative, we use the portion of the table headed by NEGATIVE z- SCORES

   - We read the table "backwards." In the body of the table find a cumulative area value that is close to our indicated area of .20; the closest value in the table is .2005

   - Find the row and column headings that label the area value of .2005
     The row is headed by –0.8; the column is headed by .04

   - Combine these two results to find $z = -0.84$

Copyright ©2014 Pearson Education, Inc.

EXAMPLE 2: Finding Values in a Non-standard Normal Distribution

Heights of American adult women are normally distributed with mean 63.6" and standard deviation 2.5".
Heights of American adult men are also normally distributed with mean 69.0" and standard deviation 2.8".

1.   What is the probability that one randomly selected American adult woman is taller than 5'5"?

- Begin by converting 5'5" to inches: 1' = 12" so 5'5" = 5(12)+5 = 65"

- Draw a representative normal distribution with the appropriate mean and standard deviation and the location of 65".

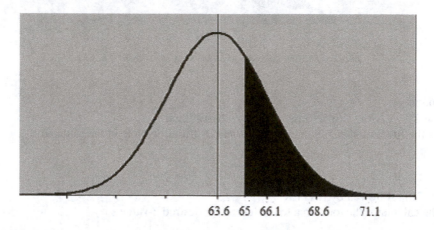

a. Using technology
- Choose 2nd VARS 2:normalcdf(followed by ENTER

- Type: 65, 99999, 63.6, 2.5) followed by ENTER

  Tell the calculator:

  o   the left boundary of the shaded region;   65

  o   the right boundary of the shaded region (since there is no actual right boundary of the shaded, choose a very large number that is many standard deviations away from the mean)   99999

  o   the mean of the distribution 63.6

  o   the standard deviation of the distribution   2.5

- Read .2877

Copyright ©2014 Pearson Education, Inc.

b. Using Table A2:

- We will find the area to the left of the shaded region and subtract from 1.

    o Convert $x = 65$ to the corresponding z- score:

    o $z = \dfrac{x - \mu}{\sigma} = \dfrac{65 - 63.6}{2.5} = 0.56$

- Go to the table and find the cumulative area value that corresponds to $z = 0.56$

    o Go down the z- column to find 0.5.

    o Go over to the column headed by .06 (note that $0.56 = 0.50 + 0.06$).

    o Read .7123 (this corresponds to the total area to the left of $z = 0.56$).

    o Now subtract from 1 to get $1 - .7123 = .2877$ (this corresponds to the total area to the right of $z = 0.56$).

2. What is the probability that the mean height of 12 randomly selected women is greater than 5'5"?

We will apply the Central Limit Theorem to respond to this question.

- Find the mean of the means
    o The mean of the means is equal to the mean of the original population

    o $\mu_{\bar{x}} = \mu = 63.6$

- Find the standard deviation of the means
    o The standard deviation of the means is found using the formula $\sigma_{\bar{x}} = \dfrac{\sigma}{\sqrt{n}}$

    o Substitute to get $\sigma_{\bar{x}} = \dfrac{2.5}{\sqrt{12}} = .72$

- Using technology we find normalcdf(65,9999,63.6,0.72)=.0259

3. The bottom 5% of American adult men are considered unusually short. The top 5% of American adult men are considered unusually tall. What are the heights that separate these unusual heights from the rest of the population of American adult men?

Begin with a appropriately labeled normal distribution.

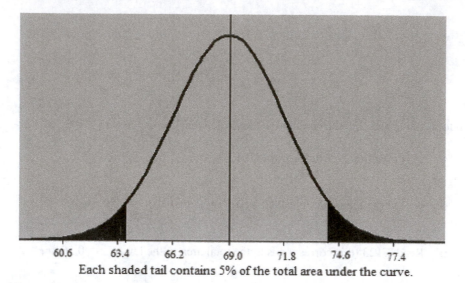

60.6   63.4   66.2   69.0   71.8   74.6   77.4

Each shaded tail contains 5% of the total area under the curve.

a. Using technology:

To find the cut-off height for the short men:

- Choose 2nd VARS 3:invNorm(followed by ENTER

- Type: .05,69,2.8)
  Tell the calculator:
      the total area to the left of the desired value, .05
      the mean of the normal distribution 69
      the standard deviation of the normal distribution 2.8

- Read: 64.4"
  Convert this to feet and inches:
  64.4 = 60+4.4 = 5' 4.4"

- CONCLUSION: All American adult men who are shorter than 5'4.4" are unusually short.

Copyright ©2014 Pearson Education, Inc.

To find the cut-off height for the tall men:

- Now choose 2nd VARS invNorm(followed by ENTER

- Type .95, 69, 2.8)
  Tell the calculator:
  Total area to the left of the desired value: .95.
  The mean of the normal distribution: 69.
  The standard deviation of the normal distribution: 2.8.

- Read 73.6"
  Convert this to feet and inches.
  73.6 = 72 + 1.6 = 6'1.6"

- CONCLUSION: All American adult men who are taller than 6'1.6" are unusually tall.

b. Using Table A2

For the height on the short side

- Look on the NEGATIVE z SCORES Section of Table A2

- Find the area value of 0.05 in the body of the table (your table may have a note at the bottom of the page identifying this special z- score)

  o Notice that 0.05 falls between 0.0505 and 0.0495

  o Find the z- value that is between the z- values for these two numbers

  o Read $z = -1.645$

- Now convert this z- value to a corresponding x- value for this distribution.
  o $x = \mu + z \cdot \sigma$
  o Substitute: $x = 69 + (-1.645)(2.8) = 64.394 \rightarrow 64.4$

To find the height on the high side we can use the symmetry of the normal distribution.

- The x- value separating the bottom 5% from the rest of the heights is 64.4

- 64.4 is 4.6 units below the mean (69 - 64.4 = 4.6)

- The x- value separating the top 5% from the rest of the heights will be same number of units *above* the mean (because of the symmetry of the distribution).

- $x = 69 + 4.6 = 73.6$

Copyright ©2014 Pearson Education, Inc.

EXAMPLE 3: Using the Normal Distribution to Approximate the Binomial Distribution

In a certain county in Florida, 23% of registered voters regularly turn out for school board elections. What is the probability that in a group of 1000 randomly selected registered voters from that county between 200 and 250 of them vote in an upcoming school board election?

First we confirm that this is a legitimate binomial distribution.

- We have a fixed number of trials: n = 1000 voters are surveyed.

- There are two outcomes:
    - success = vote in the election
    - failure = do not vote in the election

- The trials are independent: one person voting does change the probability of another person voting.

- The probability of success is the same from trial to trial: $p = 0.23$ and so $q = 1 - p = 1 - .23 = .77$.

Second, compute the mean and standard deviation for this distribution using the formulas for the binomial distribution.

- $\mu = np = 1000(.23) = 230$

- $\sigma = \sqrt{npq} = \sqrt{1000*.23*.77} = 13.3$

Next, draw an appropriate normal distribution using continuity correction values for both boundaries of the shaded region.

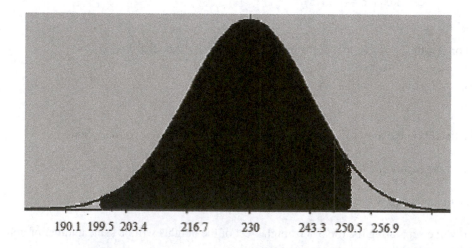

190.1 199.5 203.4   216.7   230   243.3 250.5 256.9

Finally, using technology, we compute $normalcdf(199.5, 250.5, 230, 13.3) = .9275... \rightarrow .93$

Copyright ©2014 Pearson Education, Inc.

VOCABULARY AND SYMBOL CHECK

1. Name four characteristics of the graph of the normal distribution.

2. For the standard normal distribution $\mu = $ _____ and $\sigma = $ _____.

3. The symbol for the mean of the means is _____.

4. The formula for the mean of the means is _____.

5. The symbol for the standard error of the means is _____.

6. The formula for the standard error of the means is_____.

7. Explain what an *unbiased estimator* of a population parameter is. Explain what a *biased estimator* of a population parameter is.

8. Name three examples of sample statistics that are unbiased estimators of population parameters.

9. Name three examples of sample statistics that are biased estimators of population parameters.

10. Name three different ways to determine if a data set is normally distributed.

Copyright ©2014 Pearson Education, Inc.

SHORT ANSWER

1. In a standard normal distribution find $P(1.2 < z < 2.3)$.

2. In a standard normal distribution find $P(z > 0.23)$.

3. In a normal distribution with $\mu = 12,\ \sigma = 0.8$ find $P(10 < x < 13)$.

4. In a normal distribution with $\mu = 150,\ \sigma = 12.5$ find $P(x < 145)$.

5. In a standard normal distribution find the z-value that separates the bottom 15% of the data from the top 85% of the data.

Copyright ©2014 Pearson Education, Inc.

6. In a normal distribution with $\mu = 52.1, \sigma = 7.5$ find the x-value that separates the top 12% of the data from the rest of the data.

7. In a normal distribution with $\mu = 14$, $\sigma = 2$. find the mean of the means for samples of size 25.

8. In a normal distribution with $\mu = 12$, $\sigma = 0.75$, find the standard error of the means for samples of size 64.

9. TRUE or FALSE: If the original population is normal or nearly normal, then the distribution of the sample means will be normal for any size sample. Defend your answer

10. TRUE or FALSE: The total area under the normal distribution curve depends on the value of the mean.

Copyright ©2014 Pearson Education, Inc.

Name: Date:
Instructor: Section:

11. Label the tick marks on this normal curve to correspond to a standard normal distribution.

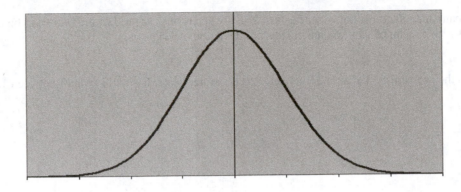

12. Label the tick marks on this normal curve to correspond to a distribution with $\mu = 15$ and $\sigma = 1.5$.

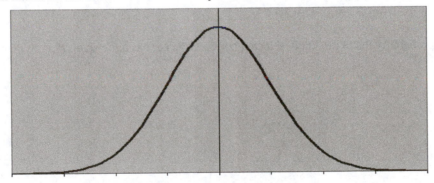

Copyright ©2014 Pearson Education, Inc.

Name: _____   Date: _____

Instructor: _____   Section: _____

PRACTICE PROBLEMS

Your company manufactures hot water heaters. The life spans of your product are known to be normally distributed with a mean of 13 years and a standard deviation of 1.5 years.

1. What is the probability that a randomly selected hot water heater has a life span of between 12 and 15 years?

2. What is the probability that a randomly selected hot water heater has a life span of more than 15 years?

Copyright ©2014 Pearson Education, Inc.

3. What is the probability that the mean life span in a group of 10 randomly selected hot water heaters is between 12 and 15 years?

4. You want to set the warranty on your product so that you have to replace at most 5% of the hot water heaters that you sell. How many years should you claim on your warranty? (Round answer to nearest whole number of years.)

CHAPTER 7: ESTIMATES AND SAMPLE SIZES

EXAMPLE 1: Construct a Confidence Interval for a Population Proportion.

In a December 2012 poll of 1002 American adults, 802 thought that if nothing is done to reduce global warming in the future, it would be a serious problem for the United States. (this category included both "not serious" or "don't know" responses). (Data from GfK Roper Public Affairs and Media.) Construct a 95% confidence interval for the proportion of American adults who describe this problem as "serious".

First, we must verify that the necessary conditions have been satisfied.

- Polling methods used by the GfK Roper Public and Media organization result in samples that can be considered simple random samples.

- Conditions of a binomial experiment have been satisfied:

  o There are a fixed number of trials, 1002 interviewees.

  o The trials are independent because one person's response does not affect the probability of a particular response from any other person.

  o There are two categories of outcome, in this case, "serious" or "not serious."

  o The probability that a person responds that this issue is "serious" remains constant from person to person.

- The 802 "successes" and the 200 "failures" are both greater than 5.

Second, we find the critical $z_{\alpha/2}$ value that corresponds to a 95% confidence level.

- Confidence level of 95% means that $\alpha = 1 - .95 = .05$ and $\alpha/2 = .05/2 = .025$.

- Using either Table A-2 or technology we find $z_{\alpha/2} = \pm 1.96$.

Third, compute the margin of error $E = z_{\alpha/2}\sqrt{\dfrac{\hat{p}\hat{q}}{n}}$.

- Find the values of the variables.

  o $\hat{p} = \dfrac{800}{1002} = .7984$

  o $\hat{q} = 1 - \hat{p} = 1 - .7984 = .2016$

  o $n = 1002$

- Now substitute to get $E = 1.96\sqrt{\dfrac{.7984 * .2016}{1002}} = 0.0248...$which rounds to 0.025.

Copyright © 2014 Pearson Education, Inc.

Fourth, construct the interval .

- $\hat{p} - E = .7984 - .025 = .7734$ (This is the lower limit of the confidence interval.)

- $\hat{p} + E = .7984 + .025 = .8234$ (This is the upper limit of the confidence interval.)

- $.773 < p < .823$ (This is the confidence interval.)

To construct this interval using technology

- Choose STAT   TESTS A:1-PropZInt   followed by ENTER

- Follow the menu prompts and enter

    - x: 800
    - n: 1002
    - C-Level: .95

- Read (.77356, .82324)
  Recall interval notation from algebra and translate to get $.773 < p < .823$.

Notice that the results obtained from the different methods are identical.

Finally, interpret the result: I am 95% confident that the actual proportion of American adults who are "totally happy" is between 77.3% and 82.3%.

EXAMPLE 2: Construct a Confidence Interval for a Population Mean

In a simple random sample of 51 community college statistics students, the mean number of college credits completed was $\overline{x} = 50.2$ with standard deviation $s = 8.3$. Construct a 98% confidence interval for the mean number of college credits completed by community college statistics students.

First, verify that the necessary conditions have been satisfied.

- The sample is a simple random sample

- $n > 30$

Second, decide on the appropriate distribution

- σ is unknown (we have only sample data) and $n > 30$

- use the student-t distribution

Copyright © 2014 Pearson Education, Inc.

Using the critical value approach:

Third, find the critical value $t_{\alpha/2}$ on Table A-3

- degrees of freedom = $n - 1 = 51 - 1 = 50$

- confidence level = 98% so $\alpha = 1 - .98 = .02 \rightarrow \alpha/2 = .02/2 = .01$

- read down the left column to 50 (degrees of freedom), then over to the column headed by 0.01 (area in one tail) to read $t_{\alpha/2} = \pm 2.403$

Fourth, compute the margin of error $E = t_{\alpha/2}\left(\dfrac{s}{\sqrt{n}}\right)$

- $t_{\alpha/2} = \pm 2.403$

- $s = 8.3$ (This is given information.)

- $n = 51$ (This is given information.)

- Substitute into the error formula to get $E = 2.403\left(\dfrac{8.3}{\sqrt{51}}\right) = 2.79...$ which rounds to 2.8.

Next, construct the interval.

- $\bar{x} - E = 50.2 - 2.8 = 47.4$ (This is the lower limit of the interval.)

- $\bar{x} + E = 50.2 + 2.8 = 53.0$ (This is the upper limit of the interval.)

- $47.4 < \mu < 53.0$ (This is the interval.)

Using technology to construct the interval:

- Choose STAT   TESTS   8:Tinterval   followed by ENTER

- Choose Stats

- Follow the menu prompts

- Read: (47.407, 52.993) which translates to $47.4 < \mu < 53.0$

Finally, interpret the result. I am 98% confident that for the population of community college statistics students, the mean number of completed college credits lies between 47.4 hours and 53.0 hours.

Copyright © 2014 Pearson Education, Inc.

EXAMPLE 3: Construct a 99% Confidence Interval for a Population Standard Deviation

A simple random sample of 25 ball bearings is selected. The mean diameter of the ball bearings is $\bar{x} = 3mm$ with a standard deviation of $s = .05mm$. Previous investigation of these ball bearings suggests that the diameters' measures are normally distributed. Construct a 95% confidence interval for the standard deviation of the diameters of the ball bearings in this population.

First, verify that the conditions have been satisfied.
- The sample is a simple random sample.
- The population appears to be normal.

Second, find the critical values for χ^2 using either Table A-4 or technology. Using the table:
- Degrees of freedom = $n - 1 = 25 - 1 = 24$.

- 95% confidence level mean $\alpha = 1 - .95 = .05 \rightarrow \alpha/2 = .05/2 = .025$.

- for χ_R^2
  - Go down the left column to 24 (degrees of freedom) and
  - over to the column headed by .025 (area to the right of the critical value) .
  - Find $\chi_R^2 = 39.364$.

- for χ_L^2
  - Go down the left column to 24 (degrees of freedom) and
  - over to the column headed by 0.975 (total area to the right of the critical value).
  - Find $\chi_L^2 = 12.401$.

Third, compute the lower and upper limits of the confidence interval and write the interval for this variance.

- lower limit: $\dfrac{(n-1)s^2}{\chi_R^2} = \dfrac{24(.05)^2}{39.364} = .0015... \rightarrow .002$

- upper limit: $\dfrac{(n-1)s^2}{\chi_L^2} = \dfrac{24(.05)^2}{12.401} = .0048... \rightarrow .005$

- $.002 < \sigma^2 < .005$

Fourth, since we want a confidence interval for the standard deviation we must take the square root of all three portions of this interval: $\sqrt{.002} < \sqrt{\sigma^2} < \sqrt{.005} \rightarrow .045 < \sigma < .071$

Finally, interpret this interval. I am 95% confident that the actual standard deviation of the diameters of this population of ball bearing lies in between 0.045mm and 0.071mm.

Copyright © 2014 Pearson Education, Inc.

Name: Date:
Instructor: Section:

VOCABULARY AND SYMBOL CHECK

1. Symbol for right tail critical value for the chi-squared distribution is _____.

2. Symbol for left tail critical value for the student-t distribution is _____.

3. α is found by _____.

4. The symbol for margin of error in a confidence interval is _____.

5. A point estimate is _____.

6. A confidence interval is _____.

7. A confidence level of 95% means _____.

8. The symbol for sample proportion is _____.

9. A critical value is _____.

10. The margin of error for a confidence interval means _____.

Copyright © 2014 Pearson Education, Inc.

SHORT ANSWER

1. The best point estimate for the population mean is the _____.

2. The basic shape of the student-t distribution is _____.

3. The basic shape of the chi-squared distribution is _____.

4. If we increase the size of our sample but keep the same confidence level, what happens to the width of the confidence interval? Why?

5. If we increase the level of confidence but keep the same sample size, what happens to the width of the confidence interval? Why?

6. TRUE or FALSE. The best point estimate for a population parameter is the corresponding sample statistic. Defend your answer.

7. TRUE or FALSE. In the confidence interval $12 < \mu < 16$ the margin of error is $E = 4$. Defend your answer.

8. TRUE or FALSE. To construct a confidence interval for the mean of a population that appears to be very skewed with $n = 10$ and unknown σ, the critical value would be $t_{a/2}$. Defend your answer.

9. TRUE or FALSE. The higher the confidence level is, the wider the confidence interval will be. Defend your answer.

10. We want to construct a confidence interval for a population mean. We know the value of σ. The population is known to be normally distributed. What distribution should we use? Explain why.

11. We want to construct a confidence interval for a population mean. We do not know the value of σ. Our sample size is 125. The population is not normally distributed. What distribution should we use? Explain why.

12. We want to construct a confidence interval for a population mean. We know the value of σ. The population is not normally distributed. Our sample size is 49. What distribution should we use? Explain why.

13. We want to construct a confidence interval for a population proportion. The conditions for a binomial distribution are satisfied and our sample has 10 successes and 35 failures. What distribution should we use? Explain why.

14. We want to construct a confidence interval for a population standard deviation. The population is known to be normally distributed. What distribution should we use? Explain why.

PRACTICE PROBLEMS

1. What is the critical value $z_{\alpha/2}$ for a 98% level of confidence?

2. What is χ_R^2 for a sample of size $n = 23$ at the 95% level of confidence?

3. What is the margin of error for the confidence interval $0.123 < p < .147$?

4. What is the value of $\hat{p}$ for the confidence interval $0.123 < p < .147$?

5. What is the value of $t_{\alpha/2}$ for a sample of size $n = 39$ at the 90% level of confidence?

6. What is the margin of error for a 95% confidence interval when $\hat{p} = 0.52$ and $n = 50$?

7. What is the lower limit of the 99% confidence interval for a population variance when $s = 3, n = 20$?

8. 50 randomly selected statistics students were asked if they actually read their statistics textbook. 35 responded that they do not read the textbook. What is the value of $\hat{p}$ for this sample?

9. A certain randomly selected sample of 125 registered voters showed that 20% of them voted in the most recent school board election. How many of these voters actually voted in that election?

10. We want to construct a confidence interval for a population mean. We want to be 95% confident with a margin of error of 1.5. We know that $\sigma = 10.5$. How large should our sample be?

Copyright © 2014 Pearson Education, Inc.

CHALLENGER:

In a CNBC-AP poll conducted in April 2010, 44% of those surveyed said that marijuana and alcohol should have the same level of governmental regulation. The poll has a margin of error of $\pm 4.3\%$. There were 1001 American adults in the survey. What is the confidence level for this poll?

Copyright © 2014 Pearson Education, Inc.

CHAPTER 8: HYPOTHESIS TESTING

EXAMPLE 1: Testing a Claim about a Population Proportion

Support for the U.S. death penalty for convicted murderers remains stable at 63% (Gallup poll results 2012). Americans support for the death penalty has leveled out in the low 60s in recent years, after several years in which support was diminishing. In a survey of 150 randomly selected 18-20 year olds, 75 said they were in favor of the death penalty for those convicted of murder.

Does this sample appear to come from a population with a lower proportion in favor of the death penalty? Use $\alpha = .05$.

First, confirm that the conditions for a binomial distribution are satisfied.

- Fixed number of trials: $n = 125$, number of people in the sample

- There are exactly two outcomes:
    - Success = in favor of capital punishment for murder convictions
    - Failure = opposed to capital punishment for murder convictions

- The trials are independent: one 18-20 year-old being in favor of the death penalty does not change the probability of the next 18-20 year-old being favor of the death penalty.

- The probability of success remains the same from trial to trial: $p = .63$ so $q = 1 - p = .37$.

- $np = (125)(.63) = 78.75 \geq 5$
  $nq = (125)(.37) = 46.25 \geq 5$
  Both values are greater than 5 so we may use the normal approximation to the binomial.

Second, work with the claim:

- Claim in words: The proportion in this population is less than 46%.
  The words "lower proportion" translate to "is less than."

- Translate the claim into symbols: $p < .63$
  This is a claim about a population proportion so remember to use the correct symbol, p.

- Opposite of the claim: $p \geq .37$
  This is the statement that must be true if the claim is false.

Next, create the null and alternative hypotheses:

- $H_0 : p = .63$
  This is always a statement of equality.

- $H_1 : p < .63$
  The alternative hypothesis is always a strict inequality.

  Notice that this is the statement of the claim.

  The inequality sign "points" to the left so we will conduct a left-tailed test.

Copyright ©2014 Pearson Education, Inc.

Traditional method:

- Compute the test statistic:

    o $z = \dfrac{\hat{p} - p}{\sqrt{\dfrac{pq}{n}}}$

    o $\hat{p} = \dfrac{x}{n} = \dfrac{75}{150} = .50$

    o p, q, n are all found in the first step above.

    o Substitute and compute: $z = \dfrac{.50 - .63}{\sqrt{\dfrac{.63 * .37}{150}}} = -3.30$.

- Use the normal distribution, shade the critical region, locate the critical value.

- Find the critical value, z_α, using Table A2 or *invNorm* in the calculator.

    This is a left-tailed test so our critical value will be negative. $z_\alpha = -1.645$

- Plot the test statistic on the horizontal axis of your normal distribution

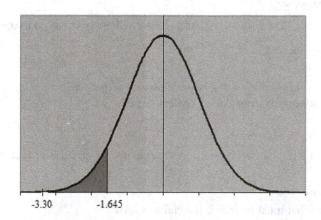

-3.30   -1.645

- Is the test statistic in the critical region?
    o Yes, -3.30 is to the left of -1.645 so it lies in the left-tail critical region.
    o When the test statistic falls in the critical region, we reject the null hypothesis.

- CONCLUSION: There is not sufficient evidence to reject the claim (as stated in the alternative hypothesis) that these American adults come from a population where the percent of people supporting the death penalty for those convicted of murder is less than 46%.

Copyright ©2014 Pearson Education, Inc.

Using the *P*- value method:

- The initial steps are the same as for the traditional method up to and including computing the test statistic.

- Plot the test statistic on the horizontal axis of your normal distribution.

- Using either technology or Table A2, find the area to the left of the test statistic. We go to the left because the alternative hypothesis told us we were conducting a left-tailed test. This is our *P*- value.
    - *normalcdf* $(-9999, -3.30) = .00048...$

    - $P = .00048 < .05$

- Because $P < \alpha$ we reject the null hypothesis.

- CONCLUSION: There is not sufficient evidence to reject the claim that these 18-20 year olds come from a population where the percent of people supporting the death penalty for those convicted of murder is less than 63%.

EXAMPLE 2: Testing a claim about a Mean with Unknown σ

The mean age of cars driven by commuting college students is seven years. The Dean of Students at FSCJ (Florida State College at Jacksonville) claims that this is an accurate statement for his students. A random sample of 31 cars in the East Student Parking Lot on campus showed a mean age of 8.1 years with a standard deviation of 5.1 years. Test the Dean's claim at the $\alpha = .05$ level of significance.

Traditional Method

- First, confirm that all necessary conditions are satisfied

    - We have a simple random sample

    - $n = 31$ so we have a large enough sample that we don't need to worry about the distribution of the population of ages of all student cars (though we suspect this would probably be a normal distribution).

    - We do not know σ so we will use a student- t distribution.

- Second, work with the claim

    - Claim in words: mean age of student cars at FSCJ is seven years.
      The words "accurate statement" translate to: mean age is the same as the general population

    - Claim in symbols: $\mu = 7$
      This is a claim about a population mean so use the appropriate symbol, m.

    - Opposite of the claim: $\mu \neq 7$
      This is what must be true if the claim is false.

Copyright ©2014 Pearson Education, Inc.

- Next, create the null and alternative hypotheses.

  o $H_0 : \mu = 7$

    The null is always a statement of equality.
    Note that this is the statement of the claim.

  o $H_1 : \mu \neq 7$

    The alternative is always a strict inequality.
    Because $\neq$ means $<$ or $>$ we conduct a two-tailed test.

- Now, compute the test statistic.

  o $t = \dfrac{\overline{x} - \mu_{\overline{x}}}{\dfrac{s}{\sqrt{n}}}$

  o $\overline{x} = 8.1,\ s = 5.1,\ n = 31$ are all given.

    $\mu_{\overline{x}} = \mu =$ the value stated in the claim$=7$.

  o Substitute and compute.

    $z = \dfrac{8.1 - 7}{\dfrac{5.1}{\sqrt{31}}} = 1.200...$

- Find the critical t values from Table A3.

  o Degrees of freedom $= n - 1 = 31 - 1 = 30$ so go down the left-hand column to d.f. = 30

  o This is a two-tailed test with $\alpha = .05$ so move over to the column headed by *area in two tails* *0.05*. (Note that the alternative heading for this column is *area on one tail 0.025* which is what you have in each tail when you create two symmetric critical regions.)

  o Read: $t = 2.042$

    Because this is a two-tailed test in a symmetric distribution, we know we have $t = \pm 2.042$

Copyright ©2014 Pearson Education, Inc.

- Using the student-t distribution, shade the critical regions, locate and label the critical values, and plot the test statistic on the horizontal axis of the distribution.

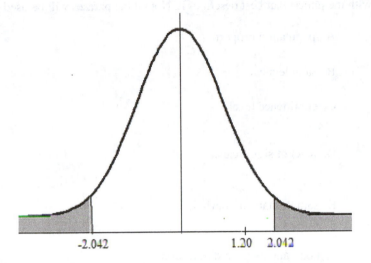

- Is the test statistic in the critical region?
    - No, the test statistic value of 1.200 is not in the shaded right tail critical region.

    - Do not reject the null hypothesis.

- CONCLUSION: There is not sufficient evidence to reject the Dean's claim that the mean age of cars driven by commuting students at his college is 7 years.

Using the P- value method.

- The steps up to and including computing the test statistic are the same.

- Use the *t-test* option in the TI calculators to read $P = 0.239$
    - Choose STAT TESTS 2:T-Test followed by ENTER

    - Fill in the information that is requested

    - Choose CALCULATE

- Now observe that $P > \alpha$.

- CONCLUSION: We do not reject the null. There is not sufficient evidence to reject the Dean's claim that the mean age of cars driven by commuting students at his college is 7 years.

Copyright ©2014 Pearson Education, Inc.

Name: Date:
Instructor: Section:

VOCABULARY AND SYMBOL CHECK

Match each symbol with the phrase that best describes it. Not all the phrases will be used.

_____1. H_0 A. population proportion

_____2. H_1 B. sample mean

_____3. α C. confidence level

_____4. p D. level of significance

_____5. $\hat{p}$ E. sample standard deviation

_____6. s   F. population standard deviation

_____7. σ G. critical value for two-tailed claim about a proportion

_____8. μ H. critical value for a claim about variance

_____9. $\bar{x}$ I. sample size

_____10. χ_R^2 J. population mean

_____11. $z_{\alpha/2}$ K. critical value for a two-tailed claim about a mean with unknown s

_____12. $t_{\alpha/2}$ L. null hypothesis

_____13. n M. alternative hypothesis

N. sample proportion

O. test statistic for claim about a population proportion

P. test statistic for claim about a population mean

Copyright ©2014 Pearson Education, Inc.

8-6

SHORT ANSWER

1. Find the critical z-value(s) for a right tailed test with $\alpha = .02$. Assume a normal population.

2. Find the critical t- value(s) for a two tailed test with $n = 12$, $\alpha = .05$. Assume a normal population.

3. Find χ_R^2 for a right-tail test with $n = 20$, $\alpha = .01$. Assume a normal population.

4. Find χ_L^2 and χ_R^2 for a two-tailed test with $n = 25$, $\alpha = .10$

5. Compute the test statistic for a claim about a population proportion given $x = 12$, $n = 25$, $p = .35$. Assume a normal population.

6. Compute the test statistic for a claim about a population mean given $\overline{x} = 143$, $\mu = 150$, $\sigma = 5.6$, $n = 10$. Assume a normal population.

7. Compute the test statistic for a claim about a population standard deviation given $n = 23$, $s = 1.4$, $\sigma = 1.5$. Assume a normal population.

8. Express each claim in symbolic form.

   a.   At least half the students at this school carpool to campus.

   b.   The average number of sick days taken annually by individual faculty members at this school is 4.

   c.   The mean annual rainfall for this county is no more than 42".

   d.   More than 25% of the injuries caused by fireworks occur to hands.

   e.   The standard deviation of the widths of ball bearings produced by my new machine is at most 0.01".

   f.   89% of people using the Sleep Number mattress experience improved sleep quality.

Copyright ©2014 Pearson Education, Inc.

9. Convert each of these claims into a null hypothesis, H_0, and an alternative hypothesis, H_1.

    a. $p = .45$

         H_0 :

         H_1 :

    b. $\sigma < 2$

         H_0 :

         H_1 :

    c. $\mu \neq 50$

         H_0 :

         H_1 :

    d. $\mu \leq 12$

         H_0 :

         H_1 :

    e. $\sigma \geq 15$

         H_0 :

         H_1 :

Copyright ©2014 Pearson Education, Inc.

10. Find the P- value for each of the following.

    a.   test statistic $z = 1.76$ in a left-tail test

    b.   test statistic $z = 1.76$ in a right-tail test

    c.   test statistic $z = -0.57$ in a two-tail test

    d.   test statistic $z = 2.67$ in a right-tail test

11. What is your conclusion in each of the following scenarios?

    a.   Claim: $p = .34$, P- value $= 0.014$, $\alpha = .05$

    b.   Claim: $\mu < 6$; test statistic: $z = -1.75$; $\alpha = .01$

    c.   Claim: $\sigma \geq 65$; test statistic: $\chi^2 = 22.6$; $\alpha = .05$, $n = 22$

Copyright ©2014 Pearson Education, Inc.

PRACTICE PROBLEMS

1. In 2007 the National Fire Protection Association reported that 22% of all fireworks injuries affected the eyes. The administrator of a local hospital claims that in his community the percentage is actually higher. His ER reported 40 eye injuries in the 154 patients who presented with fireworks injuries. Test the administrator's claim at the $\alpha = .05$ level of significance.

2. The average life span of the largemouth bass fish is 16 years. An ichthyologist in Georgia is concerned that oil contamination of the feeding grounds for this fish in his area has resulted in a shortened life span for the bass. He has recorded the ages of 25 bass. His sample shows a mean age of 13.2 years with a standard deviation of 2.3 years. Test his claim of a shortened life span at the $\alpha = .05$ level of significance.

3. A retired statistics professor has recorded final exam results for decades. The mean final exam score for the population of her students is $\mu = 82.4$ with a standard deviation of $\sigma = 6.5$. In the last year, her standard deviation seems to have changed. She bases this on a random sample of 25 students whose final exam scores had a mean of $\bar{x} = 80$ with a standard deviation of 4.2. Test the professor's claim that the current standard deviation is different. Use $\alpha = .05$.

4. The professor in the previous problem contends that the mean final exam scores for her students has not changed. Test this claim using the data in the previous problem.

Copyright ©2014 Pearson Education, Inc.

CHAPTER 9: INFERENCES FROM TWO SAMPLES

EXAMPLE 1: Inferences about Two Proportions

56 randomly selected male students at a commuter college were asked if they lived with their parents. 24 responded yes. 51 randomly selected female students at the same commuter college were asked if they lived with their parents. 21 responded yes. The Dean of Students believes that more males live independently of their parents. Test the Dean's assumption. Use $\alpha = .05$.

First, confirm that all requirements for this testing are satisfied.

- The samples are independent
    - One sample is all males
    - The other sample is all females
    - There was no pairing of any of the participants in the two groups

- The number of successes (yes) and failures (no) is greater than five in both groups.

Second, determine the claim and identify the null and alternative hypotheses.

- Claim: more males living independently translates to the statement that the proportion of female students who live with their parents is less than the proportion of male students who live with their parents. We let the males represent Population 1 and the females represent Population 2.

    In symbols: $p_1 > p_2$

- $H_0: p_1 = p_2$ (Notice: this means $p_1 - p_2 = 0$.)

    $H_1: p_1 > p_2$ (This tells us we will have a right-tailed test.)

Third, assign values to all the necessary variables in each sample. All these formulas or descriptions are defined in your text.

- Population 1 is the males so we have:

    - n_1 = size of male sample = 56

    - x_1 = successes in male sample = 24

    - $\hat{p}_1$ = male sample proportion = $\dfrac{x_1}{n_1} = \dfrac{24}{56} = .429$

    - $\hat{q}_1 = 1 - \hat{p}_1 = .571$

- Population 2 is the females so we have:

    - n_2 = size of female sample = 51

    - x_2 = successes in female sample = 21

○　$\hat{p}_2$ = female sample proportion = $\dfrac{x_2}{n_2} = \dfrac{21}{51} = .412$

○　$\hat{q}_2 = 1 - \hat{p}_2 = 1 - .412 = .588$

- Pooled sample proportion = $\bar{p} = \dfrac{x_1 + x_2}{n_1 + n_2} = \dfrac{21 + 24}{51 + 56} = .421$

- $\bar{q} = 1 - \bar{p} = 1 - .421 = .579$

Now, compute the test statistic. We will need this value for either the P- value test or a critical-value test.

- $z = \dfrac{(\hat{p}_1 - \hat{p}_2) - (p_1 - p_2)}{\sqrt{\dfrac{\bar{p} \cdot \bar{q}}{n_1} + \dfrac{\bar{p} \cdot \bar{q}}{n_2}}}$　where $p_1 - p_2 = 0$ (This assumption is in the null hypothesis.)

Substitute to get $z = \dfrac{(.429 - .412) - 0}{\sqrt{\dfrac{(.421)(.579)}{56} + \dfrac{(.421)(.579)}{51}}} = .178$.

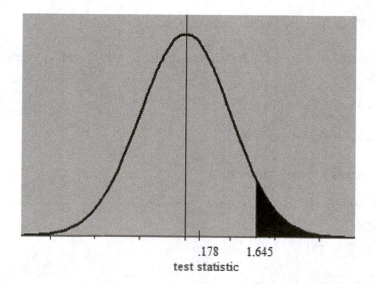

Using the traditional method we see that the test statistic is not in the critical region so we do not reject the null hypothesis that the population proportions are equal. There is not sufficient evidence to support the claim that a greater proportion of male students live independently.

Using the *P*-value method we compute the area to the right of the test statistic to get $P = .430$. This is greater than $\alpha = .05$ so we do not reject the null hypotheses that the population proportions are equal. There is not sufficient evidence to support the claim that a greater proportion of male students live independently.

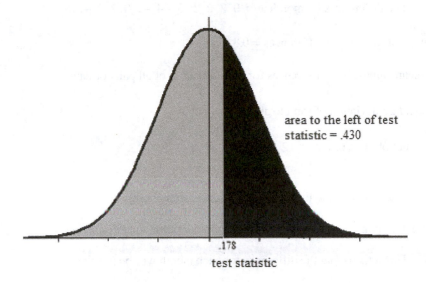

area to the left of test statistic = .430

.178
test statistic

EXAMPLE 2: Inferences from Dependent Samples

The Director at a local duplicate bridge club claims that there are many more director calls for rules violations on the day of a full moon. (The director at a duplicate bridge game is sort of like the referee at a boxing match!) She has collected a year's worth of data using the day of the new moon and the day of the full moon.

MONTH	J	F	M	A	M	J	J	A	S	O	N	D
NEW MOON	2	2	3	4	3	6	3	4	1	0	2	3
FULL MOON	2	4	5	6	6	2	1	4	3	2	1	7

Use the $\alpha = .05$ level of significance to test the Director's claim that there are more calls on days of a full moon.

First, check to see that requirements for the test procedure are satisfied.

- The sample data are dependent. The two sets of data were gathered from the same sample of bridge players.

- The samples are simple random samples.

- We may reasonably assume that the population here is essentially normally distributed. There are no extreme departures from normality. The normal quantile plot is nearly linear which also suggests a nearly normal distribution.

Copyright ©2014 Pearson Education, Inc.

Second, identify the variables we will be working with.

- $d =$ individual differences between the two values in a single matched pair
    - We compute the differences by finding (full moon calls – new moon calls).

    - The differences by month are: 0 2 2 2 3 –4 –2 0 2 2 –1 4.

- $\bar{d} =$ mean of the sample differences = 0.8

- $\mu_d =$ mean value of the differences for the population of all pairs of data

- $s_d =$ standard deviation of sample differences = 2.3

- $n =$ number of pairs of data = 12

Third, identify the claim and create the null and alternative hypotheses.

- The director claims that there are more calls on the full moon days.

    - This tells us that the difference for each month will be positive.

    - Therefore the claim translates to $u_d > 0$

    - This statement will become the alternative hypothesis

    - The direction of the inequality tells us we have a right tail critical region.

- The null hypothesis is always a statement of equality. The null should state that there is no significant difference between the number of full moon calls and the number of new moon calls so we say $m_d = 0$

Next, determine the appropriate distribution. We do not know σ, the population standard deviation , so we use the student-t distribution.

Now, compute the test statistic.

- $t = \dfrac{\bar{d} - \mu_d}{\dfrac{s_d}{\sqrt{n}}}$

- using the values identified above, substitute and compute

$$t = \frac{.8 - 0}{\dfrac{2.3}{\sqrt{12}}} = 1.205$$

Using the traditional method, we draw the distribution, locate the right tail critical region, find the critical value from Table A3 and plot the test statistic on the horizontal axis.

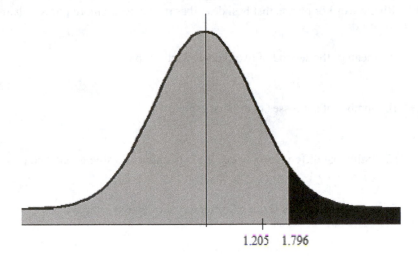

1.205 1.796

- The test statistic is not in the critical region so we do not reject the null hypothesis.

- CONCLUSION: There is not sufficient evidence to support the Director's claim that there are more director calls on the day of a full moon. Apparently, she just feels that she is being unduly harassed on those days!

Using the P- value method we use the TI83/84 calculator to conduct a T-Test

- Choose STAT   TESTS   2:T-Test followed by ENTER

- Using the Stats option, follow the menu prompts to enter the relevant information.

  - $\mu_0 = 0$
  - $\bar{x} = \bar{d} = .8$
  - $s_x = s_d = 2.3$
  - $n = 12$
  - At the final prompt choose $\mu :> \mu_0$. This creates the right tail test we want.

- The screen shows a P- value of 0.1267.

  $P > .05$ so we do not reject the null hypothesis.

  Notice that the test statistic value, $t = 1.20490491$, is slightly different from our computed value. This is due entirely to round-off error. If we were to compute the test statistic using four or five significant digits for all our variables, our result would match the result on the screen.

- CONCLUSION: There is not sufficient evidence to support the Director's claim that there are more director calls on the day of a full moon. Apparently, she just feels that she is being unduly harassed on those days!

Copyright ©2014 Pearson Education, Inc.

VOCABULARY AND SYMBOL CHECK

Match each symbol with the word or phrase that best describes it. There are more phrases than symbols so some phrases will not be used.

_____1. $\hat{p}_1$ A. mean of the second of two independent samples

_____2. n_1 B. number of successes in the first sample

_____3. x_1 C. individual difference between the two values in a single matched pair

_____4. $\overline{p}$ D. mean value of the set of all differences for paired sample data

_____5. $\overline{d}$ E. standard deviation of the differences for paired sample data

_____6. μ_d F. larger of two sample variances taken from two independent populations

_____7. s_d G. smaller of two sample variances

_____8. s_1^2 H. proportion from the first sample

_____9. d I. size of the first sample

_____10. $\overline{x}_2$ J. mean value for the population of differences of all paired data

K. pooled proportion for two samples

L. standard deviation of the population of all possible sample means $\overline{x}$

M. standard score

Copyright ©2014 Pearson Education, Inc.

SHORT ANSWER

1. The formulas in this chapter are more involved than most. Try your hand (or your calculator) at evaluating each of the following:

   a. $z = \dfrac{(\hat{p}_1 - \hat{p}_2) - (p_1 - p_2)}{\sqrt{\dfrac{\overline{pq}}{n_1} + \dfrac{\overline{pq}}{n_2}}}$ with $\hat{p}_1 = .366$, $\hat{p}_2 = .300$, $p_1 = p_2$, $\overline{p} = .337$, $n_1 = 500$, $n_2 = 400$

   b. $E = z_{\alpha/2}\sqrt{\dfrac{\hat{p}_1\hat{q}_1}{n_1} + \dfrac{\hat{p}_2\hat{q}_2}{n_2}}$ with $\hat{p}_1 = .892$, $\hat{p}_2 = .752$, $n_1 = 54$, $n_2 = 25$, $\alpha = .01$

   c. $t = \dfrac{(\overline{x}_1 - \overline{x}_2) - (\mu_1 - \mu_2)}{\sqrt{\dfrac{s_1^2}{n_1} + \dfrac{s_2^2}{n_2}}}$ with $\overline{x}_1 = 15.6$, $\overline{x}_2 = 17.0$, $\mu_1 = \mu_2$, $s_1 = .2$, $s_2 = .13$, $n_1 = 12$, $n_2 = 18$

   d. $E = z_{\alpha/2}\sqrt{\dfrac{\sigma_1^2}{n_1} + \dfrac{\sigma_2^2}{n_2}}$ with $\alpha = .05$, $\sigma_1 = 2.2$, $\sigma_2 = 3.1$, $n_1 = 22$, $n_2 = 25$

   e. $t = \dfrac{\overline{d} - \mu_d}{\dfrac{s_d}{\sqrt{n}}}$ with $\overline{d} = 7.8$, $\mu_d = 0$, $s_d = 0.6$, $n = 16$

Copyright ©2014 Pearson Education, Inc.
9-7

2. Explain what is meant by *independent samples*.

3. Explain what is meant by *dependent samples*.

4. Under what circumstances would a confidence interval test about the means of independent samples lead us to reject the null hypothesis?

5. Under what circumstances would a hypothesis test about a claim about the means from two independent samples lead us to reject the null hypothesis?

6. Under what circumstance would we reject the null hypothesis when we are conducting a *P*- value test for a claim about two proportions?

Copyright ©2014 Pearson Education, Inc.

7. We want to test a claim about two population proportions. We want to use the methods of this chapter. What conditions must be satisfied?

8. We want to test a claim about two independent population means. We want to use the methods of this chapter. What conditions must be satisfied?

9. We want to test a claim about the mean of the differences from dependent samples. We want to use the methods of this chapter. What conditions must be satisfied?

10. We want to test a claim about two population standard deviations or variances. We want to use the methods of this chapter. What conditions must be satisfied?

Copyright ©2014 Pearson Education, Inc.

PRACTICE PROBLEMS

1. Biologists have identified two subspecies of largemouth bass swimming in US waters, the Florida largemouth bass and the Northern largemouth bass. The mean weight for Florida species is 11 pounds, mean weight for Northern species is 7 pounds. On two recent fishing trips you have recorded the weights of fish you have captured and released. Use this data to test the biologists claim that the mean weight of the Florida bass is different from the mean weight of the Northern bass. Use $\alpha = .05$. Try to show your results in both the traditional method and the P- value method.

   Florida bass weights, in pounds: 5 6 5 8 12 11 10 7 16 13
   Northern bass weights, in pounds: 5 8 3 4 7 10 12 9 6 10

Name: _____   Date: _____

Instructor: _____   Section: _____

2. A randomly selected group of 35 adult American men was asked if they owned a handgun. 20 said yes.
   A randomly selected group of 40 adult American women was asked if they owned handgun. 21 said yes.
   Does there appear to be a significant difference between the proportion of adult men who own handguns and adult women who own handguns? Use $\alpha = .05$

Copyright ©2014 Pearson Education, Inc.

3. A random sample of 15 employees of a large manufacturing installation was selected. Each employee was asked to report the number of sick days he/she claimed on Friday and then on Wednesday of the previous calendar year. Use this information to test the employer's claim that more employees call in sick on Friday than on Wednesday. Use $\alpha = .05$

	#1	#2	#3	#4	#5	#6	#7	#8	#9	#10	#11	#12	#13	#14	#15
FRI	2	3	1	1	6	2	4	0	5	2	3	3	4	1	0
WED	1	1	0	2	1	3	0	1	4	2	3	1	2	1	1

Copyright ©2014 Pearson Education, Inc.

Chapter 10: CORRELATION AND REGRESSION

EXAMPLE 1: Linear Correlation and the Regression Line Equation

12 randomly selected women were asked to record their height in inches and their shoe size. The results are shown in the following table. Use $\alpha = .05$

Ht, in inches	67	66	64	64	67	64	68	65	68	65	66	61
Shoe Size	10	7	7	9	8	8.5	8.5	8.5	9	8	7.5	6

1. What pattern, if any, does a scatter plot of these ordered pairs suggest?

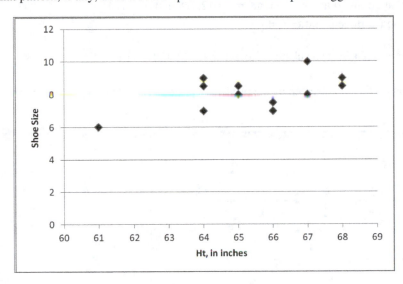

The pattern in the plot looks fairly linear. There do not appear to be any outliers or influential points.

2. We use the calculator to find the value of the linear correlation coefficient, r.
   - Enter the x values (height in inches) into some convenient list
     Enter the y values (shoe size) into some other convenient list

   - Choose STAT   TESTS
     For TI-83: Choose E:LinRegTTest   followed by ENTER
     For TI-84: Choose F:LinRegTTest   followed by ENTER

   - Follow the prompts on the menu list
     o Enter the location of your x and y values by naming the appropriate list
     o Choose β & $\rho : \neq 0$
     o Scroll down to CALCULATE followed by ENTER

   - Read: P- value = .045853713 and $r = .58469$.

     o Since the P- value is less than the significance level of .05, we reject the null hypothesis that $r = 0$ and we conclude that there is sufficient evidence to support the claim of linear correlation between height and shoe size.

Copyright ©2014 Pearson Education, Inc.

     o   Use Table A6 to interpret r

        ❖ Using $\alpha = .05$ and $n = 12$ we read a critical value of .576.

        ❖ Since $|r| > .576$, we reject the null hypothesis that $\rho = 0$ and we conclude that there is sufficient evidence to support the claim of linear correlation between height and shoe size.

3.   Construct the linear regression equation.

- From the same screen that got us ρ and r, we read $\alpha = -12.428$ and $b = .314$. We use the round-off rule for three significant digits.

- Use the model equation $y = \alpha + bx$ and substitute to get $\hat{y} = -12.428 + .314x$.

4. Graph the regression equation on the same screen as the original scatter plot to see how well the regression line fits the data.

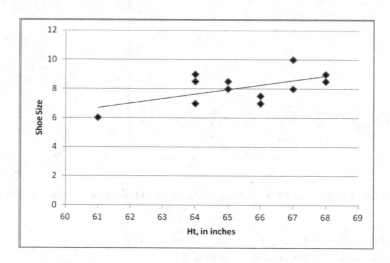

5.   Because we have determined that there is significant linear correlation between these variables, we can use the regression equation to predict the shoe size of a woman of a given height.

- What size shoe do you predict for a woman who is 5'5" tall?

     o   Convert height to inches: 5'5" tall = 65"

     o   Substitute $x = 65$ into the regression equation to find the predicted y-value
$\hat{y} = -12.428 + .314x$ so $\hat{y} = -12.428 + .314(65) = 7.98 \approx 8$

- What shoe size do you predict for a woman who is 6'2"tall?

    A height of 6 feet is well outside the range of x-values contained in our original sample so using the regression equation to make a prediction for this height is not recommended.

    However, if we forge ahead and substitute we get: $\hat{y} = -12.428 + .313(74) = 10.808 \approx 11$ which does not appear to be an unreasonable result.

Copyright ©2014 Pearson Education, Inc.

6. Lastly, we look at the residual plot

    •   Create a new table of values using all the original x values paired with the residual for each
        y-value. To find the residual, find the difference between the actual y value and the predicted
        y-value

        o   Actual value comes from the original data set

        o   Predicted value comes from $\hat{y} = -12.428 + .314x$

Ht, in inches	67	66	64	64	67	64	68	65	68	65	66	61
Predicted Shoe Size	8.61	8.30	7.67	7.67	8.61	7.67	8.92	7.98	8.92	7.98	8.30	6.73
Residual	1.39	-1.30	-0.67	1.33	-0.61	0.83	-0.42	0.52	0.08	0.02	-0.80	-0.73

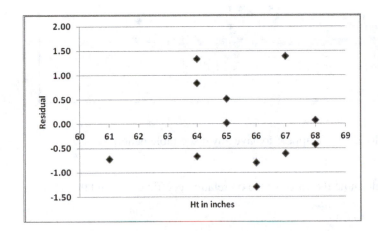

Since the scatter plot for the residuals shows no obvious pattern, we are reassured that the regression
equation is a good model for the sample data

Copyright ©2014 Pearson Education, Inc.

EXAMPLE 2: Linear Correlation and the Regression Line

12 women were randomly selected. They were asked their age and the number of US states they had visited. The responses are shown in the table below. Use $\alpha = .05$

Age	23	20	22	20	23	19	20	25	20	32	35	21
# States	5	12	8	16	8	8	10	15	20	6	14	7

1. Look at the scatter plot to determine any pattern.

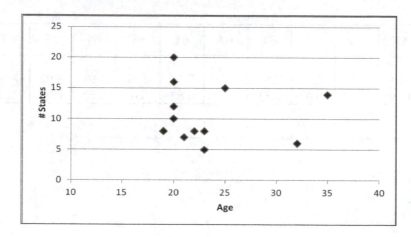

This plot does not appear to have any discernible pattern.

2. Go to the calculator to determine the value of the correlation coefficient r and the P- value.

We go to the LinRegTTest to find

- $r = -.0733$

  ○ From Table A6 we read the critical value of .576

  ○ $|r| < .576$ so we do not reject the null hypothesis that $\rho = 0$

  ○ CONCLUSION: There is no significant linear correlation between the age of the respondents and the number of states they have visited.

- Using P- value = .821

  ○ Since $P > \alpha$ we do not reject the null hypothesis that $\rho = 0$

  ○ CONCLUSION: There is no significant linear correlation between the age of the respondents and the number of states they have visited.

Copyright ©2014 Pearson Education, Inc.

3. Construct the linear regression equation and show it graphed along with the scatter plot to see how well or not well the line fits the plot.

- Using the values from LinRegTTest we have

  - $a = 12.307$
  - $b = -.0667$
  - $\hat{y} = 12.307 - .0667x$

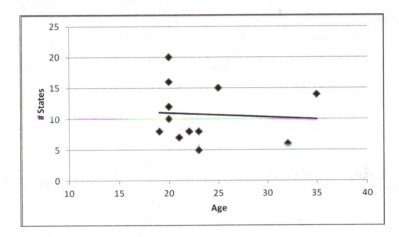

   Once again we observe no linear pattern and no fit between the regression line and the original data points.

4. Predict the number of states that a 30 year old woman has visited.

- Since we have do not have significant linear correlation, our best predictor for the number of states visited is the mean y-value from the original data points.

- Using the calculator and 1:VarStats on our list of original y-values we read $\bar{y} = 10.75$.
   Our best estimate of the number of states visited by a 30-year old woman is 11 states. (Round the 10.75 to the nearest whole number since she cannot visit a fractional part of a state.)

5. Look at the residual plot. First, create the table of residual values.

Age	23	20	22	20	23	19	20	25	20	32	35	21
Predicted # of States	10.77	10.97	10.84	10.97	10.77	11.04	10.97	10.64	10.97	10.17	9.97	10.91
Residual	-5.77	1.03	-2.84	5.03	-2.77	-3.04	-.97	4.36	9.03	-4.17	4.03	-3.91

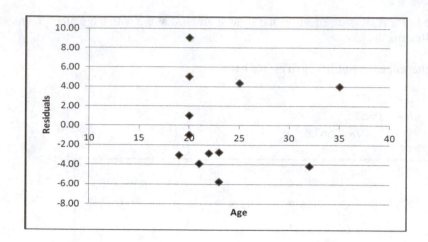

The plot shows a "thinning" of the points suggesting that the regression equation is not a good model.

VOCABULARY AND SYMBOL CHECK

1. What is a residual?

2. What is an influential point?

3. Name four characteristics of the sample linear correlation coefficient r.

4. What is the symbol for the population linear correlation coefficient?

5. What is the null hypothesis statement in a test for linear correlation?

Copyright ©2014 Pearson Education, Inc.

SHORT ANSWER

1. TRUE or FALSE: Correlation implies causality. Defend your answer.

2. TRUE or FALSE: The regression equation is always the best predictor of a y value for a given value of x. Defend your answer.

3. TRUE or FALSE: The regression line always passes through at least one point in the original data set. Defend your answer.

4. TRUE or FALSE: The correlation coefficient r can be used to measure strength of both linear and non-linear correlation. Defend your answer.

5. TRUE or FALSE: The residual value is the difference between the observed y- value and the predicted y- value, $\hat{y}$.

6. The regression line equation for a set of data is given by $\hat{y} = 2.3x + 5$. $n = 10$. The value of $\bar{y}$ for this same data set is 10.1. Use $\alpha = .05$

    a.   If the linear correlation coefficient for this data is $r = .521$, what is the best y-value for $x = 5$?

    b.   If the linear correlation coefficient for this data is $r = .972$, what is the best predicted y-value for $x = 5$?

7. What is the critical value for the linear correlation coefficient r for a sample of size $n = 15$ with $\alpha = .01$?

8. The linear correlation coefficient for a set of paired variables (x, y) is $r = .897$. What proportion of the variation in y can be explained by the linear relationship between x and y?

9. Identify two different conditions under which the regression line should not be used to make predictions.

PRACTICE PROBLEMS

1. A video game player insists that the longer he plays a certain computer game, the higher his scores
   are. The table shows the total number of minutes played and the high score (in thousands of points) achieved
   after each 5-minute interval. Use $\alpha = .05$.

# Min	5	10	15	20	25	30	35	40	45	50	55	60
Score	48	53.3	101.9	72.5	121.5	146	196.1	118.5	150.5	80.7	36	64.8

a.   Draw a scatter plot for this data. Do you see a pattern?

b.   Find the linear correlation coefficient, r. Use r and Table A6 to determine if there is significant linear
     correlation .

c.   Find the P- value. Use P and α to determine if there is significant linear correlation.

d.   Based on your responses to parts (b) and (c), what game score do you predict after 47 minutes of play?

Copyright ©2014 Pearson Education, Inc.

2. The heights and weights of 15 randomly selected adult men are shown in the table. Use $\alpha = .05$.

Ht.	73	70	73	72	68	72	68	70	68	67	72	70	65	74	70
Wt.	185	155	195	164	145	185	170	150	180	175	175	147	140	210	200

a. Draw a scatter plot for this data. Do you see a pattern?

b. Find the linear correlation coefficient, r. Use r and Table A-6 to determine if there is significant linear correlation.

c. Find the P-value. Use P and α to determine if there is significant linear correlation.

d. Based on your responses to parts (b) and (c), what weight do you predict for a man who is 66" tall?

Name: Date:

Instructor: Section:

3. 12 randomly selected adults were asked how many US states they had visited and how many foreign countries they had visited. The responses are shown in the table. Is there significant linear correlation between these variables?

# States, x	20	27	8	8	32	7	3	10	32	25	20	8
# countries, y	0	5	0	7	5	5	0	5	10	4	4	8

Copyright ©2014 Pearson Education, Inc.

CHAPTER 11: GOODNESS-OF-FIT AND CONTINGENCY TABLES

EXAMPLE 1: Goodness–of-Fit Test

The Human Resources Department of a certain corporation wants to know if some days of the week experienced greater absenteeism than other days. 100 sick day reports are randomly selected. The day of the week of the absences is recorded. Test the claim that sick days occur with the same frequency on every day of the week. Use $\alpha = .05$

Day	Mon	Tue	Wed	Thur	Fri
Number Sick	22	15	17	16	30

First, conduct the requirement check.

- The data have been randomly selected. We trust the Human Resources Department to have done this.

- The sample data has frequency counts for each of the different categories. There are absences reported for each day of the week.

- The expected frequencies for each day of the week are all at least 5. We expect the same frequency for each day. With 100 absences being examined, we expect 20 absences on each day.

Second, identify the variables that you will be working with.

- Observed and Expected frequencies can easily be compared in a table

O	22	15	17	16	30
E	20	20	20	20	20

- The associate expected probabilities are:

  p_1 = probability that an absence occurs on Monday

  p_2 = probability that an absence occurs on Tuesday

  p_3 = probability that an absence occurs on Wednesday

  p_4 = probabliity that an absence occurs on Thursday

  p_5 = probability that an absence occurs on Friday

Third, determine the statements of the claim and the hypotheses.

- The claim is that the absences occur with the same frequency on every day of the week.

- This translates to $p_1 = p_2 = p_3 = p_4 = p_5$

- $H_0: p_1 = p_2 = p_3 = p_4 = p_5$ (Remember, the null hypothesis is always a statement of equality.)

  H_1: at least one of these probabilities is different from the others.

Copyright ©2014 Pearson Education, Inc.

Next, compute the test statistic

$$\chi^2 = \sum \frac{(O-E)^2}{E}$$ Once again, this is most efficiently done in a table

	Observed, O	Expected, E	$O - E$	$(O - E)^2$	$\dfrac{(O - E)^2}{E}$
Monday	22	20	2	4	0.2
Tuesday	15	20	-5	25	1.25
Wednesday	17	20	-3	9	0.45
Thursday	16	20	-4	16	0.8
Friday	30	20	10	100	5
				Sum of entries in last column	**7.7**

The final sum is the value of the test statistic: $\chi^2 = 7.7$

Traditional Method:

- Use Table A-4 to find the critical value for χ^2

    o $\alpha = .05$ in the right tail

    o   degrees of freedom = number of categories − 1 = 5-1 = 4

    o   read $\chi^2 = 9.488$

- The test statistic is less than the critical value, so it does not fall in the critical region; therefore we do not reject the null hypothesis.

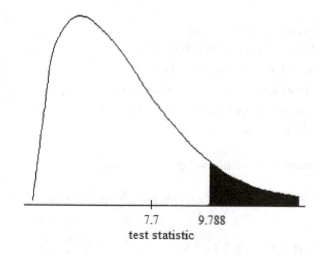

7.7   9.788
test statistic

- CONCLUSION: There is not sufficient evidence to reject the claim that the absences occurred with the same frequency on every day of the week.

Copyright ©2014 Pearson Education, Inc.

Using *P*-value Method

- Enter the observed values into L1 (or some other convenient list)

- Enter the expected values into L2 (or some other convenient list)

- Choose STAT   TESTS   D: χ^2 GOF-TEST followed by ENTER

- Confirm that the calculator has the correct location of your observed and expected values.

- Enter the degrees of freedom: number of categories $- 1 = 5 - 1 = 4$.

- Choose CALCULATE followed by ENTER.

- Read:
    - $\chi^2 = 7.7$ (Notice, this agrees with our computed value of the test statistic.)

    - $P = .1032067217$
      Because $P > \alpha$ we do not reject the null hypothesis

- CONCLUSION: There is not sufficient evidence to reject the claim that the absences occurred with the same frequency on every day of the week.

EXAMPLE 2: Test of Independence

A random sample of students at a commuter college was asked if they lived at home with their parents, rented an apartment or owned their own home. The results are shown in the table below sorted by gender. Test the claim that living accommodations are independent of gender. Use $\alpha = .05$

Table of Observed Values

	Live with Parent	Rent Apartment	Own Home	Row Totals
Male	20	22	10	**52**
Female	21	22	5	**48**
Column Totals	**41**	**44**	**15**	**100**

As usual, we begin with the requirement check.

- Sample data was randomly selected.

- Sample data has been collected and displayed as frequency counts in a contingency table.

- Every cell in the table has an expected value of at least 5.

Use the formula $E = \dfrac{(\text{row total})(\text{column total})}{\text{grand total}}$ to find the expected value in each cell.

Copyright ©2014 Pearson Education, Inc.

Table of Expected Values

	Live with Parents	Rent Apartment	Own home
Male	$\frac{52*41}{100} = 21.3$	$\frac{52*44}{100} = 22.88$	$\frac{52*15}{100} = 7.8$
Female	$\frac{48*41}{100} = 19.68$	$\frac{48*44}{100} = 21.12$	$\frac{48*15}{100} = 7.2$

Identify the null and alternative hypotheses:

- H_0 : Living accommodations are indepedent of gender

  H_1 : Living accommodations are dependent on gender

Traditional Method:

- Compute the test statistic using the formula $\chi^2 = \sum \frac{(O-E)^2}{E}$

  Use the two tables above to find values; begin in the upper left cell and move across the rows to get:

$$\chi^2 = \frac{(20-21.3)^2}{21.3} + \frac{(22-22.88)^2}{22.88} + \frac{(10-7.8)^2}{7.8} + \frac{(21-19.68)^2}{19.68} + \frac{(22-21.12)}{21.12} + \frac{(5-7.2)^2}{7.2} = 1.53$$

- Go to Table A4 to find the critical value

  - Degrees of freedom = (number of rows - 1)(number of columns - 1) = (2 - 1)(3 - 1) = 2

  - This is always a right tail test so go to the column headed by .05 (area to right of critical value)

  - Read: 5.991

- Observe that the test statistic does not fall in the critical region so we do not reject the null hypothesis.

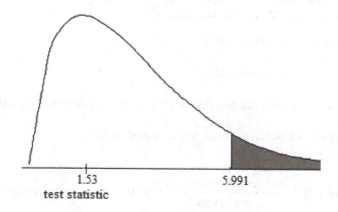

1.53
test statistic

5.991

- CONCLUSION: Living accommodations appear to be independent of gender.

Copyright ©2014 Pearson Education, Inc.

Using the *P*-value method

- Enter the observed values into your calculator in a matrix

  o Choose 2nd x^{-1} EDIT 1:[A] followed by ENTER

  o Define the size of the matrix as 2x3 followed by ENTER

  o Move through each cell of the matrix entering the observed value from our table above

- Go to the STAT menu

  o Choose TESTS C: χ^2-TEST followed by ENTER.

  o Your data is already entered into [A].

  o Ignore [B] – this is the table of expected values – the calculator will do this computation automatically.

  o Scroll down to CALCULATE and hit ENTER.

  o Read

  χ^2-TEST
  χ^2 =1.533510527
  p=.4645178637
  df=2

  o Notice that the test statistic χ^2 value and degrees of freedom value are the same as ones we calculated.

  o Read *P*-value = .4645. $P > \alpha$ so we do not reject the null hypothesis.

- CONCLUSION: There is not sufficient evidence to reject the claim that living accommodations are independent of gender.

Copyright ©2014 Pearson Education, Inc.

Name: Date:
Instructor: Section:

VOCABULARY AND SYMBOL CHECK

1. Define *observed frequency*.

2. Define *expected frequency*

3. The formula for the test statistic in both the goodness-of-fit test and the contingency table test for independence is _____.

4. The formula for expected value in any cell of a contingency table is _____.

5. In a hypothesis test for independence of the row and column variable in a contingency table, the null hypothesis is _____.

Copyright ©2014 Pearson Education, Inc.
11-6

6. The formula for degrees of freedom in a contingency table test for independence is _____.

7. The claim in a test of homogenicity is _____.

8. McNemar's test is used for _____.

9. A goodness-of-fit test is used for _____.

10. Define *contingency table*.

Copyright ©2014 Pearson Education, Inc.

Name: Date:
Instructor: Section:

SHORT ANSWER

1. Use this contingency table to respond to the questions that follow:

	In Favor of Amendment	Opposed to Amendment	Undecided
Male	12	16	20
Female	18	9	15

a. What are the row totals for this table?

b. What are the column totals for this table?

c. What is the grand total for this table?

d. What are the expected values for each of the cells in this contingency table?

e. What is the χ^2 critical value for this test? (Use $\alpha = .05$)

f. What is the test statistic for this table?

g. What is the null hypothesis for a test of independence for this table?

h. What is the P- value for this test?

Copyright ©2014 Pearson Education, Inc.

2. Use the table shown here to respond to the questions that follow:

Style of Vehicle	Mini-van	Pick-up Truck	Convertible	SUV	Sedan
Number	15	8	7	12	20

  a. If we want to test the claim that these styles of vehicles occur with the same frequency, what should the expected values for each cell be?

  b. What is the χ^2 critical value for this test? Use $\alpha = .05$.

  c. What is the null hypothesis for a goodness-of-fit test for this data?

  d. What is the P- value for this test?

3. TRUE or FALSE: Tests of independence with a contingency table may be two-tailed. Defend your answer.

4. TRUE or FALSE: When conducting a hypothesis test for independence between row and column variables in a contingency table, we must guarantee that the data come from an essentially normally distribution population. Defend your answer.

5. TRUE or FALSE: The expression "If P is low, the null must go" tells us to reject the null hypothesis when the P- value is sufficiently small. Defend your answer.

Copyright ©2014 Pearson Education, Inc.

Name: Date:
Instructor: Section:

PRACTICE PROBLEMS

1. A random sample of students from a commuter college was asked about their employment status. The results are recorded in the table below. Does the employment status appear to be dependent on the age of the student? Use $\alpha = .05$

	No Job	Part-time Job	Fulltime Job
Under 21	12	29	4
21 or over	11	35	21

Copyright ©2014 Pearson Education, Inc.

2. A random sample of students at a commuter college were surveyed and asked about their employment status. The results are recorded in the table below. Test the claim that the frequencies are evenly distributed throughout the categories. Use $\alpha = .05$.

# Hours per week	0-9 hours	10-19 hours	20-29 hours	30-39	40+
Number of students	28	15	19	22	24

Copyright ©2014 Pearson Education, Inc.

3. 68 randomly selected people were asked to test the efficiency of a new style of game controller. The frequencies for successful completion of the test and the handedness of the subjects are recorded below. The manufacturer claims that this new device works equally well for left-handed and right-handed people. Test this claim at the $\alpha = .05$ level.

	Successful	Unsuccessful
Left-handed	12	22
Right-handed	18	16

Copyright ©2014 Pearson Education, Inc.

CHAPTER 12: ANALYSIS OF VARIANCE

EXAMPLE 1: Using One-Way ANOVA to Test Equality of Three Population Means

Randomly selected students were asked to report their age and the total number of college credits they had completed. The students were divided into three groups: teenagers (under 20 years of age), 20-somethings (20-29 years of age) and 30-somethings (30-39 years of age). Use one-way analysis of variance (ANOVA) to test the claim that the mean number of completed college credits is the same for all three populations. Use $\alpha = .05$.

Teens	24	24	28	31	28	24	72	42	30	33	31	36	53	25	54
20's	58	28	131	33	34	48	31	45	60	44	83	51	60	18	124
30's	73	39	80	45	53	64	142	124	22	84	102	140	30	53	94

First step, requirement check:

- Normal quantile plots of the 20's and 30's data are very nearly linear so we trust they come from normally distributed populations. The plot for the teen data is less convincing, but for the purposes of this example, we will trust that this population has a distribution that is close enough to normal.

- The populations must have the same standard deviations. Again, the 20's and 30's samples have nearly identical variances. The teen data has a smaller variance; however, since the sample sizes are identical and since the largest variance is less than nine times the smallest variance, ANOVA methods will be reliable.

- The samples are simple random samples gathered from a variety of classrooms.

- The samples are independent. There is no overlap of the age groups; there was no pairing of the data.

- The populations are categorized by age only.

Identify the null and alternative hypotheses.
- $H_0 : \mu_1 = \mu_2 = \mu_3$ The three population means are the same.

- $H_1 :$ at least one of the means is different from the others

Using the TI843/84 calculator

- Enter the data for the three populations into three separate lists

- Choose STAT TEST ANOVA(followed by ENTER

- After the blinking cursor on your homescreen, type in the location of the lists containing your data, followed by a comma, followed by ENTER.

- Find the P-value.

  o $P = .00278 < \alpha = .05$
  o P is small enough to reject the null hypothesis of equal means

CONCLUSION: at least one of the age groups has a population mean different from the others.

Copyright ©2014 Pearson Education, Inc.

EXAMPLE 2: Using Two-Way ANOVA to test for Interaction Effect, Row Factor Effect and Column Factor Effect.

Randomly selected students were asked how many college credit hours they had completed. Results are organized by age and gender. The results are displayed in the table below.

	Teens	20's	30's
Female	53	83	45
	31	30	73
	28	90	124
	24	53	68
	28	62	64
Male	40	60	84
	26	44	53
	30	51	58
	33	60	131
	12	35	75

Begin with the requirement check.

- As in the previous example, we are reasonably confident that this data comes from populations that are nearly normally distributed.

- The variances in the cells range from 108.2 (male teens) to 963.7 (male 30's). With such small samples we would need more variation than this to reject equal variances.

- The samples are simple random samples selected by the author in a variety of classrooms.

- The samples are independent; the respondents were not matched in any way.

- The sample values are categorized two ways: gender and age.

- All the cells have the same number of sample values. There are five sample values for each cell.

Following the directions in your text for Excel, Minitab or the TI calculator, we proceed. The display shown here is from Excel using ANOVA: Two-factor with Replication.

Source of Variation	SS	df	MS	F	P-value	F crit
Sample	136.5333	1	136.5333	0.295836	0.591522	4.259677
Columns	11097.27	2	5548.633	12.02261	0.000241	3.402826
Interaction	451.6667	2	225.8333	0.489329	0.619022	3.402826
Within	11076.4	24	461.5167			
Total	22761.87	29				

Note: The row titles in Excel are slightly different from those shown in the Minitab displays in your text. The correspondences are: Type in Minitab = Sample in Excel; Size in Minitab = Columns in Excel; Error in Minitab = Within in Excel

Copyright ©2014 Pearson Education, Inc.

Step 1: Test the null hypothesis that there is no interaction between the two factors of age and gender. The P-value for interaction is .619022 so we fail to reject the null that there is no interaction between age and gender. (Remember, if the P-value is low the null must go; otherwise, fail to reject.) It does not appear that the number of completed college credits is affected by an interaction between the age and the gender of the student.

Step 2: Because there does not appear to be any interaction between the two factors, we go next to test for effects from the row and column factors. The two hypotheses are:

H_0 : There are no effects from the row factor (i.e., the row means are equal).

H_0 : There are no effects from the column factor (i.e., the column means are equal)

For the row factor (type or sample) we have a test statistic of

$$F = \frac{MS(sample)}{MS(within)} = \frac{136.5333}{461.5167} = .295836$$ The corresponding P-value is .591522.

Because the P-value is greater than the significance level of 0.05 we fail to reject the null hypothesis and we conclude that number of completed college credits does not appear to be affected by the gender of the student.

Step 3: Compute the test statistic $F = \frac{MS(column)}{MS(within)} = \frac{5548.633}{461.5167} = 12.02261$. The corresponding

P-value is .000241 Because the P-value is less than the significance level of 0.05 we reject the null hypothesis and we conclude that number of completed college credits does appear to be affected by the age of the student.

Copyright ©2014 Pearson Education, Inc.

VOCABLARY AND SYMBOL CHECK

1. What does ANOVA stand for?

2. What does one-way analysis of variance measure?

3. What does two-way analysis of variance measure?

4. What is meant by interaction between two factors?

5. What is meant by effect of a row factor?

6. What is meant by effect of a column factor?

Copyright ©2014 Pearson Education, Inc.

SHORT ANSWER

1. Name three characteristics of the F distribution.

2. TRUE or FALSE: The critical value for F is obtained using a right-tail test. Defend your answer.

3. TRUE or FALSE: In addition to identifying that populations have different means, the ANOVA test will also identify specifically which particular population means are different. Defend your answer.

4. What feature will an interaction graph display if there is little or no interaction between the row and column variables?

5. Under what circumstance should we proceed with a test of row/column effects when conducting a two-way ANOVA test?

Copyright ©2014 Pearson Education, Inc.

PRACTICE PROBLEMS

1. The table below shows pulse rates taken from random samples of adults. The data have been sorted by age. (Data taken from data set 1 in Appendix B of the textbook.)

20-29	64	64	76	64	60	88	72	56	88	72	68	80	72	72	68	64
30-39	88	72	85	60	84	84	64	56	72	68	80	76	60	76	80	60
40-49	72	60	84	72	56	64	70	76	68	96	72	64	80	104	88	124

Use one-way analysis of variance (ANOVA) to test the claim that the three different age groups have the same mean pulse rate. Remember to begin with a requirement check.

Copyright ©2014 Pearson Education, Inc.

2. The pulse rate data can be grouped by gender as well as age.

		20-29	30-39	40-49
Male		64	88	72
		64	72	60
		76	84	84
		64	60	72
		60	84	56
		88	84	64
		72	64	70
		56	56	76
Female		88	72	68
		72	68	96
		68	80	72
		80	76	64
		72	60	80
		72	76	104
		68	80	88
		64	60	124

Use two way analysis of variance to test for an interaction effect, an effect from the row factor (gender), and an effect from the column factor (age).

Copyright ©2014 Pearson Education, Inc.

CHAPTER 13: NONPARAMETRIC STATISTICS

EXAMPLE 1: Runs Test for Randomness:

In the card game of bridge, an opening hand is generally considered to be one that has 12 or more high card points. At a recent duplicate tournament, the person sitting in the north seat claimed that he had a particularly bad spell of hands with below opening count; the machine that generated the hands claims the deals are all random. The table below shows the point count for each of north's hand at the tournament. The hands have been sorted into two categories: P = pass for non-opening hand; B = bid for opening hand. Test the player's claim that he was not given randomly dealt cards. Use $\alpha = .05$.

Points	9	7	10	5	3	12	8	11	10	20	17	11	8	8	12	11	11	10	12	7
Category	P	P	P	P	P	B	P	P	P	B	B	P	P	P	B	P	P	P	B	P

Begin with a requirement check.

- The data are arranged in the order in which the hands were dealt.

- Each value has been assigned to one of two categories: P = pass for non-opening hand; B=bid for opening hand

Next, identify the claim.

- H_0: the sequence of opening and non-opening hands is random

- H_1: the sequence of opening and non-opening hands is not random

Determine the values of all the necessary variables

- n_1 = number of P items = 15

- n_2 = number of B items = 5

- G = number of runs = 9

Since both n_1 and n_2 are less than or equal to 20 and since $\alpha = .05$, our test statistic is $G = 9$.

Go to Table A10 to find critical values for G

- using $n_1 = 15$ and $n_2 = 5$ we find critical values of 4 and 12

Our $G = 9$ value is between the two critical values so we do not reject the null hypothesis of randomness.

CONCLUSION: the sequence of opening and non-opening hands appears to be random.

EXAMPLE 2: Sign Test

The table below shows the highway MPG amounts for a sample of 6-cylinder cars. Test the claim that median MPG rating for 6-cylinder cars is more than 30.

MPG	26	29	29	27	27	25	25	30	37	32	27	31
Sign	-	-	-	-	-	-	-	0	+	+	-	+

Begin with the requirement check.

 The only requirement is that the data come from a simple random sample. We are using the data presented by the author in the Appendix to the book so we trust that the sample is random.

Identify the hypotheses:

- H_o : the median is 30   (Remember that the null hypothesis is always a statement of equality.)

- H_1 : the median is more than 30

Attach a plus or minus sign to each data value depending on whether it is above or below the claimed median value. Any data value that is equal to the claimed median value is assigned a 0.

Assign values to the appropriate variables.

- n = total number of signs (discarding any 0's) = 11

- x = number of the less frequent sign = 3   (There are 3 plus signs and 8 minus signs.)

Use the binomcdf function (access by 2nd Vars) in the TI83/84 calculator to find the P- value

- binomcdf(11,0.5,3) = 0.113

     o   Recall that the parameter list for binomcdf is n, p, x

     o $n = 11$, the value found above

     o $p = 0.5$, always use this value for the sign test

     o $x = 3$, the value found above

- $P = 0.274 > \alpha = .05$

The P- value is greater than α so do not reject the null hypothesis that the median is 30.

CONCLUSION: the evidence does not support the claim that the median MPG rating for 6-cylinder cars is more than 30.

Copyright ©2014 Pearson Education, Inc.

EXAMPLE 3: Wilcoxon Rank-Sum Test for Two Independent Samples

We want to test the claim that the median ages of female Statistics students and male Statistics students are the same. Two random samples of Statistics students were gathered. The results are shown in the table below.

Male	19	46	19	26	19	24	21	22	29	31	27	30	20	21	30	28	19
Female	35	24	21	21	38	19	19	20	37	19	18	19	46	38	19	24	21

As usual, begin with a requirement check.

- We have two independent random samples

- Both samples sizes are greater than 10.

Identify the hypotheses.

- H_0 : the two samples come from populations with equal medians

- H_1 : the median of the male population is different from the median of the female population .

Now, we must assign the ranks.
- Combine the data into one big sample, then assign each value a rank. Using a List in your calculator will make it easy for you to order the data so you can assign the ranks.

   o Put the data values in a convenient list

   o Choose STAT   2:SortA(followed by ENTER

   o At the prompt on your home screen type in the name of the list that holds your data.

   o Press ENTER

   o The data in your list has now been sorted in order from smallest to largest

Remember that in the case of "ties" for a rank, use the mean rank for the positions of the shared data values.

   o The age 19 in the combined data set occupies the 2nd through the 10th positions

   o The mean of 2, 3, 4, 5, 6, 7, 8, 9, 10 is 6 so the shared rank for the 19-year olds is 6

- The tables with the corresponding ranks are shown below.

Male	19	46	19	26	19	24	21	22	29	31	27	30	20	21	30	28	19
Rank	6	33.5	6	22	6	20	15	18	25	28	23	26.5	11.5	15	26.5	24	6

Female	35	24	21	21	38	19	19	20	37	19	18	19	46	38	19	24	21
Rank	29	20	15	15	31.5	6	6	11.5	30	6	1	6	33.5	31.5	6	20	15

Copyright ©2014 Pearson Education, Inc.

Next, assign value to the appropriate variables:

- n_1 = size of the male sample = 17

- n_2 = size of the female sample = 17

- R_1 = sum of the ranks of the male sample = 312

- R_2 = sum of the ranks of the female sample = 283

- $R = R_1 = 312$

- $\mu_R = 297.5$

  - $\mu_R = \dfrac{n_1\left(n_1 + n_2 + 1\right)}{2}$

  - Substitute the values found above to get $\mu_R = \dfrac{17(17 + 17 + 1)}{2} = 297.5$.

- $\sigma_R = 29.03$

  - $\sigma_R = \sqrt{\dfrac{n_1 n_2 (n_1 + n_2 + 1)}{12}}$

  - Substitute the values found above to get $\sigma_R = \sqrt{\dfrac{(17)(17)(17 + 17 + 1)}{12}} = 29.03$.

Now compute the test statistic

- $z = \dfrac{R - \mu_R}{\sigma_R}$

  Substitute the values found above to get $z = \dfrac{312 - 297.5}{29.03} = .499$.

- This is a two-tailed test with $\alpha = .05$ so we have critical values of ± 1.96.

- The test statistic is not in the critical region so do not reject the null hypothesis that the samples come from populations with the same median.

CONCLUSION: There is no evidence to suggest a difference between the median age of male statistics students and median age of female statistics students.

Copyright ©2014 Pearson Education, Inc.

VOCABULARY AND SYMBOL CHECK

1. Explain what is meant by parametric tests.

2. Explain what is meant by nonparametric tests.

3. What is meant by the rank of an item?

4. When is the **sign test** used?

5. When do we use the Wilcoxon signed-ranks test?

Copyright ©2014 Pearson Education, Inc.

6. When do we use the **rank correlation** test?

7. Define a **run.**

8. When do we use the Kruskal-Wallis test?

9. What is the null hypothesis in the Kruskal-Wallis test?

10. What test should we use when we want to test a claim about the proportion of two nominal data categories?

Copyright ©2014 Pearson Education, Inc.

SHORT ANSWER

1. A matched data set produced the following differences: - 2.5, 0, 1.5, - .5, 1, 0, - 2, 1, 1.5, 1. You plan to run a Wilcoxon Signed-Ranks Test. What are the corresponding ranks (remember to discard the zeroes and ignore the signs)?

2. A sequence of bridge hands created by a card dealing program was examined for suits that were 6 cards or more in length. You want to conduct a runs test for randomness on this information. In the list below, Y represents a deal in which at least one hand held a 6-card or longer suit, and N represents a deal in which no hand held a 6-card or longer suit. N N Y Y N N N Y Y N N N Y Y N Y N Y N Y N Y Y N.

    a.   How many runs are displayed in this list?

    b.   What are n_1 and n_2 for this data?

    c.   What is the test statistic for this data set?

    d.   What are the critical values for this test? Use $\alpha = .05$

    e.   What is your conclusion about the randomness of this data?

Copyright ©2014 Pearson Education, Inc.

3. What does the runs test for randomness test for?

4. Assign ranks to the following values in a data set: 22, 33, 5, 9, 8, 21, 1, 22, 14, 2, 20, 19, 2, 15, 19

5. TRUE or FALSE: 20 randomly selected teenagers are asked how many text messages they send during a 24-hour period, and then how many text messages they receive during the same 24-hour period. These two data sets are independent samples because they are asking about two different things (send and receive).

6. TRUE of FALSE: 20 randomly selected men take an agility test and record the results. 20 randomly selected women take the same test and record the results. These two data sets are independent samples because they come from two different populations, men and women.

7. TRUE or FALSE: Rank correlation can sometimes detect relationships that are not linear.

Copyright ©2014 Pearson Education, Inc.

Name: Date:
Instructor: Section:

PRACTICE PROBLEMS

1. Judges in a local cheerleading competition ranked the eight contestants as shown in the table. Do the judges appear to rank about the same or are they very different?

Contestant	Annie	Barb	Celia	Dottie	Evie	Fran	Gert	Haley
Judge 1	1	5	7	4	3	6	2	8
Judge 2	8	2	5	4	6	1	3	7

Copyright ©2014 Pearson Education, Inc.

2. The table below shows neck circumferences and weights for wild bears. Measurements were taken on anesthetized bears during the month of January. Data are taken from the Data Set 6 in your text.

Neck	16	28	31	31.5	18	30.5	17	15	15	20
Weight	80	344	416	348	140	514	114	64	60	150

    a.   Show a scatter plot for the neck/weight data. Note any pattern you observe.

    b.   Test for linear correlation between neck measurements and weight.

    c. Use a 0.05 significance level with rank correlation to test for a correlation between neck measurements and weight.

Copyright ©2014 Pearson Education, Inc.

3. Listed below are daily high temperatures recorded on 10 randomly selected days in July, 2010, in Jacksonville, Florida. According to the National Weather Service the median high temperature in Jacksonville in July is 88.7. Use the sign test with a 0.05 level of significance to test the claim that the median high temperature is 88.7.

95	87	86	90	91	82	88	89	96	94

Copyright ©2014 Pearson Education, Inc.

4. Use the data in exercise (3) above to conduct the Wilcoxon signed-ranks test.

5. Compare your results from exercises (3) and (4) above. Comment.

Copyright ©2014 Pearson Education, Inc.

CHAPTER 14: STATISTICAL PROCESS CONTROL

EXAMPLE 1: Control Chart for the Mean

Each month for one calendar year, randomly selected samples of 10 cantaloupe are weighed. The samples are chosen from the same produce market each month. The mean weights of each sample are listed below along with the corresponding range value for each sample. Construct an $\bar{x}$ chart to determine if the mean weights of the cantaloupe are statistically stable.

	Jan	Feb	Mar	Apr	May	June	July	Aug	Sept	Oct	Nov	Dec
$\bar{x}$	2.47	2.46	2.28	2.55	2.37	2.55	2.42	2.54	2.58	2.25	2.51	2.6
range	0.9	0.9	0.9	0.9	0.9	0.8	0.8	0.7	0.6	0.7	0.6	0.8

Begin by doing a requirement check.

- The mean weights are all from samples of the same size, $n = 10$.

- Examination of a quantile plot of the sample means shows that the distribution is essentially normal.

- Individual sample values are independent. The cantaloupe were chosen during different months and were randomly selected.

Find the centerline, upper control limit, and lower control limit values for the data.

- $\bar{\bar{x}}$ = mean of the means = 2.465.

- Upper control limit $= UCL = \bar{\bar{x}} + A_2\bar{R}$.

   - $\bar{R}$ = mean of the individual range values = 0.79.
   - $A_2 = 0.266$
      - from an expanded version of Table 14-2 (Control Chart Constants)
      - for $n = 12$
   - Substitute to find $UCL = 2.675$.

- Lower control limit $= LCL = \bar{\bar{x}} - A_2\bar{R}$.

   - Using the same values found above substitute and compute.

   - $LCL = 2.255$

Now, draw the plot showing the data points and horizontal lines for the centerline, UCL, and LCL.

Apply the criteria for determining when a process is out-of-control

- Is there a pattern, trend or cycle that is obviously not random?
  No.

- Is there a point lying outside the region between UCL and LCL?
  Yes, there is one point falling just below the line representing the lower control limit.

Copyright ©2014 Pearson Education, Inc.

- Are there eight consecutive points either all above or all below the centerline? No.

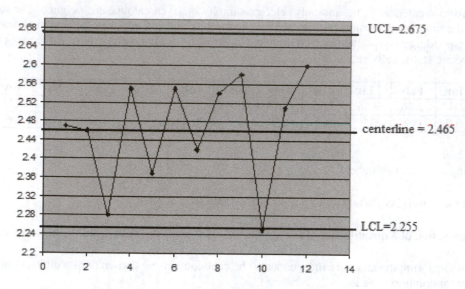

Since one of the tests fails, we conclude that the process mean values are not within statistical control, but it is a very close call.

EXAMPLE 2: Construct a Control Chart for R

Use the data in the previous problem to construct a control chart for the range.

The control chart for R has the same requirements as the control chart for $\bar{x}$ so we are satisfied that they are met.

Find the centerline, UCL and LCL.

- Centerline is $\bar{R} = .79$.

- $UCL = D_4 \bar{R}$

  o Go to an expanded version of Table 14-2 to find $D_4 = 1.717$ for n = 12

  o Substitute and compute to find $UCL = 1.356$

- $LCL = D_3 \bar{R}$

  o Go to an expanded version of Table 14-2 to find $D_3 = 0.283$

  o Substitute and compute to find LCL=0.224

Copyright ©2014 Pearson Education, Inc.

Now draw the plot showing the data points, the centerline and horizontal lines corresponding to LCL and UCL.

Apply the criteria for determining whether the process is out-of-control.

- Is there a pattern, trend or cycle that is obviously not random?
  No

- Is there a point lying outside of the region between the upper control limit and the lower control limit?
  No.

- Are there eight consecutive either all above or all below the centerline?
  No, but there are seven points above the centerline followed by an eighth that is just a tad below the centerline, so this condition is pretty close to being violated.

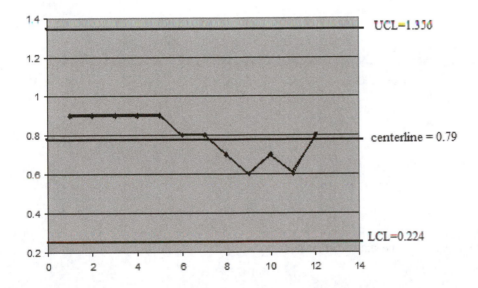

We conclude that the variance for this data is in statistical control, but the decision is a very close call.

Copyright ©2014 Pearson Education, Inc.

VOCABULARY AND SYMBOL CHECK

1. What is random variation?

2. What is assigned variation?

3. What is a run chart?

4. Explain what it means for a process to be statistically stable.

5. What is the *run rule of 8*?

Copyright ©2014 Pearson Education, Inc.

6. In a control chart for $\overline{x}$, what does $\overline{\overline{x}}$ represent?

7. What is an $\overline{x}$ chart?

8. What is a p chart?

9. What is an R chart?

10. In a p chart, what is $\overline{p}$?

Copyright ©2014 Pearson Education, Inc.

SHORT ANSWER

1. Give four examples of ways a process may not be statistically stable.

2. What are the criteria for determining if a process is not statistically stable?

3. What are the three features usually included in a control chart?

4. What are upper and lower limits of a control chart based upon?

5. Find UCL for a control chart for $\overline{x}$ given: $\overline{\overline{x}} = 72.3$, $\overline{R} = .35$, $n = 15$.

Copyright ©2014 Pearson Education, Inc.

6. Does this run chart exhibit statistical stability? Why or why not? (Data represent total withdrawals from the author's sections of statistics per semester for the past 15 consecutive semesters.)

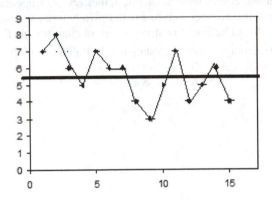

7. Find the lower control limit of a control chart for R given: $\bar{R} = 20.5$, $n = 20$

8. TRUE or FALSE. Statistical process control can be used to determine if a manufacturer's specifications are being met. Defend your answer.

9. TRUE or FALSE. When constructing a control chart, we may gather samples of various sizes as long as they are all individually large enough.

Copyright ©2014 Pearson Education, Inc.

PRACTICE PROBLEMS

1. In each of the 10 most recent and consecutive academic semesters, 50 students in the author's statistics classes were randomly chosen; the number who had withdrawn before successful completion of the course was recorded. The results are listed below. Construct a control chart for p. Based on the appearance of your chart, does the withdrawal percentage seem to be statistically stable? Why or why not?

5   6   3   4   8   2   4   4   1   2

Copyright ©2014 Pearson Education, Inc.

2. The number of people reporting problems with their cable TV systems has been recorded. The totals for each week during the past calendar year are listed below. The list is in order from the first week in January to the last week in December.

31 23 38 49 47 43 13 46 10 46 49 17 11 35 27 20 19 37 17 27 50 16 19 38 12 24
32 39 21 26 41 18 11 33 10 23 37 46 36 12 45 12 25 28 20 16 46 46 34 31 40 24

a. Construct a run chart for the 52 values. Does there appear to be a pattern suggesting that the process is not within statistical control?

b. Group the data by quarter Jan-Mar (13 weeks), Apr-June (13 weeks), July-Sept (13 weeks), Oct-Dec (13 weeks). Construct a control chart for $\bar{x}$ for these four subgroups. Is the process mean within statistical control?